7th Asia Pacific International Institute of Welding International Congress 2013

AA005197

Recent Development in Welding and Joining Technologies

Singapore
8 – 10 July 2013

ISBN: 978-1-63439-610-3

Printed from e-media with permission by:

Curran Associates, Inc.
57 Morehouse Lane
Red Hook, NY 12571

Some format issues inherent in the e-media version may also appear in this print version.

Copyright© (2013) by International Institute of Welding
All rights reserved.

Printed by Curran Associates, Inc. (2017)

For permission requests, please contact International Institute of Welding
at the address below.

International Institute of Welding
BP 51362 – VILLEPINTE
95 942 ROISSY CH DE
GAULLE Cedex
France

Phone: +33 1 4990 3608

www.iiwelding.org

Additional copies of this publication are available from:

Curran Associates, Inc.
57 Morehouse Lane
Red Hook, NY 12571 USA
Phone: 845-758-0400
Fax: 845-758-2633
Email: curran@proceedings.com
Web: www.proceedings.com

TABLE OF CONTENTS

Portevin lecture – Automation in electron beam welding is the key to quality and productivity

D. von Dobeneck, Planegg/D

Automation means that a machine does a number of chained production processes by itself without manual interference of an operator. The purpose of automation is to increase productivity, at the same time to reduce scrap rates and / or eliminate boring or tiring labour.

As the subject automation is too diversified, I will, besides a few standard solutions for the equipment, concentrate on the intelligence for automation of an electron beam. The peculiar properties and capabilities inherent only in electron beam technology are rapid and nearly inertialess direct controllability by computers. A further physical property which has been disturbing in earlier years, has now become an important carrier of automation: back scattered electrons, resulting when the beam hits material and, here in special, the weld joint.

The first step of automation is to make sure that the tool electron beam is constant and always the same in its geometrical and optical parameters. This adjustment requires skilled operators or an automated adjustment routine. The beam calibrates itself.

A typical application for automation is to weld many identical parts. It was a tiring procedure to position each part with optics and a reticule. Therefore SPS and CNC control were introduced which however needed precise tools and positioning equipment and from time to time manual input as the systems warm up and expand. During the time there were many improvements on this conventional kind of automation. However the big innovation came from introducing the information on the weld joint by back scattered electrons. The new system needs no precise tools and even drifting of the beam is no problem, as the measuring beam is at the same time the welding beam.

A number of examples for this big step in technology for automation in welding will be given.

For mass production today one piece flow is preferred. Parts may arrive at the welding station in varying positions or upside down. It is no longer necessary to rotate the part into a uniform location; the beam scans the actual position and welds accordingly with a cycle time of 6 sec.

There are numerous applications where a quality control is required before welding in order to prevent welding of misassembled or misaligned parts and thus producing scrap. The back scattered electrons signal to the CNC control whether the parts may be welded with an additional beam deflection or be rejected before welding.

Another category of work pieces carry a large number of joints, e.g. turbine vane rings with more than 100 welds on the circumference. Symmetric heat input and shrinkage from each weld has to be taken into account. This requires an online seam tracking together with a weld sequence plan. As the weld joints are 3 dimensional in space, scanning and welding have to follow the profile. With conventional semiautomatic EB welding 2 operators need 8 hours to finish 1 ring of vanes. Automated it takes 4 hours and produces much fewer defects.

Finally back scattered electrons can be used for inspection of the weld quality before the part is removed from its weld fixture and tooling. This capability allows eventual repair welding without losing a lot of additional time.

Automation in Electron Beam Welding is the Key to Improving Quality and Productivity.

1. (Title weld seam).
Mr. President, ladies and gentlemen,

2. (2 pictures automated lines).
Speaking of automation, it is generally understood that a machine does a number of chained production processes by itself without manual interference. The purpose of automation is to increase productivity, at the same time to reduce scrap rates and eliminate tiring labour. We realize this with many types of equipment supplied to our customers.
However, as this is state of the art, and as the subject automation is too diversified to deal with in one lecture, I would like to concentrate on the “intelligence” of an electron beam for automation. The peculiar property and capability inherent only in electron beam technology is the rapid and nearly inertialess direct controllability by computers in nearly real time.

Interlinked automated production lines.

3. (Diagram beam hits material).
A further physical property has now become an important carrier of automation: back scattered electrons. They are generated when the beam hits material. When the beam is scanned over a surface, this will be visualized like on a scanning microscope. When scanning across the weld joint, its position is detected.

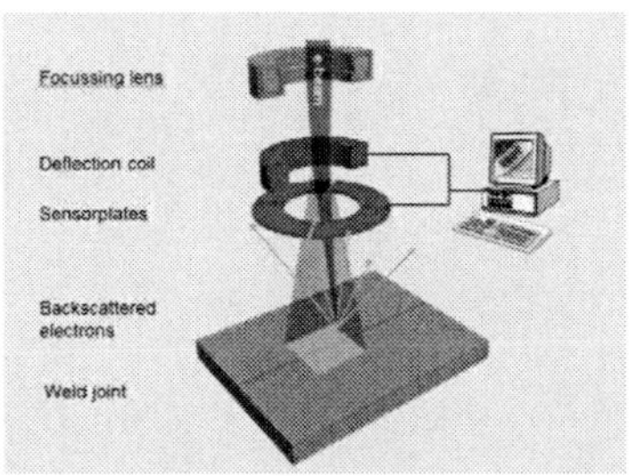

Generation and use of backscattered electrons.

4. (Deep weld, shallow weld).

The high power density allows shaping the weld as required: deep and slender or wide and shallow. The high mobility of the focus spot, initiated by electro-magnetic fields, permits manual as well as direct computer control. Transformation of the kinetic energy of the electrons into heat produce the weld pool, whereas positioning is by interpreting the back scattered electrons, generated during short intervals of welding with low beam power.

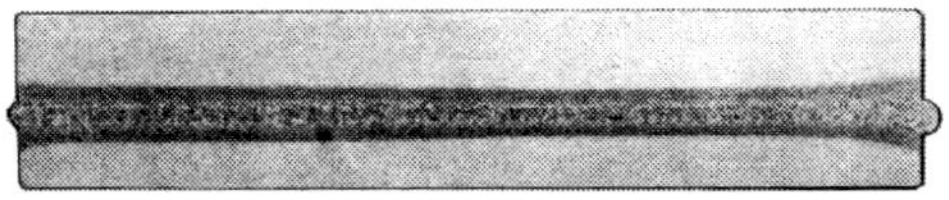

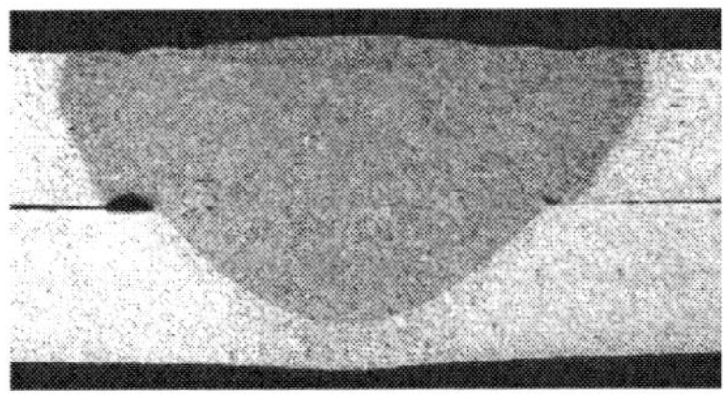

The weld shape is influenced by beam power, focus position, beam oscillation and welding speed. It may be deep and slender or shallow and wide.
left: 150 mm deep, 6 mm wide
right: 1.9 mm deep, 4 mm wide

(5. Beam Adjustment)

The first step of automation is to make sure that the thermal tool electron beam is constant and always the same in its geometrical and optical parameters. This adjustment requires skilled and experienced operators who in industry are not always available. An automated adjustment routine does exactly the same thing as an operator would do: wobbling around zero position and fix the right point. As a result the beam adjusts and calibrates itself.

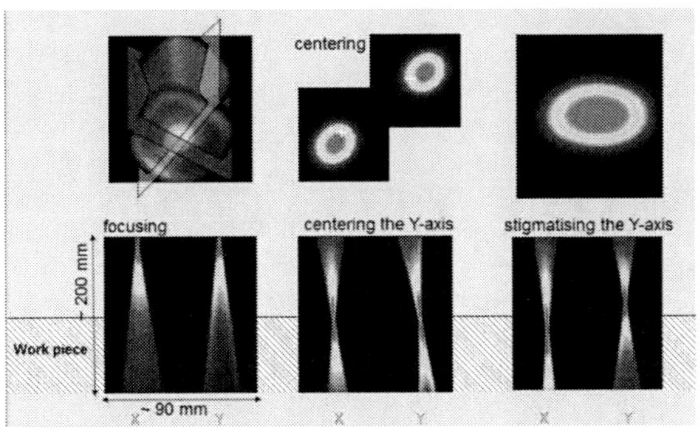

patented by pro-beam

Automated Beam Adjustment

A typical application for automation is to weld many identical parts. It was a tiring procedure to position each part with optics under a reticule. When the operator's eye got tired the amount of scrap was increasing. Therefore CNC control was introduced, which, however, needed precise tools and positioning equipment. From time to time manual input was necessary as the system warms up and expands. Therefore it was the next task to get rid of this problem.

6. (Heat exchanger 1 teaching).

Let us step by step introduce automation in the electron beam welding process. The simplest kind of a weld is linear and the operator has to position the electron beam at the start and endpoint of the joint by turning the knob of the deflection amplifier and teach the position to the computer for linear interpolation. For welding a circle he teaches 3 points and circular interpolation. For a free shape line he teaches several points and asks for spline interpolation. This semi automation is time consuming and not very precise.

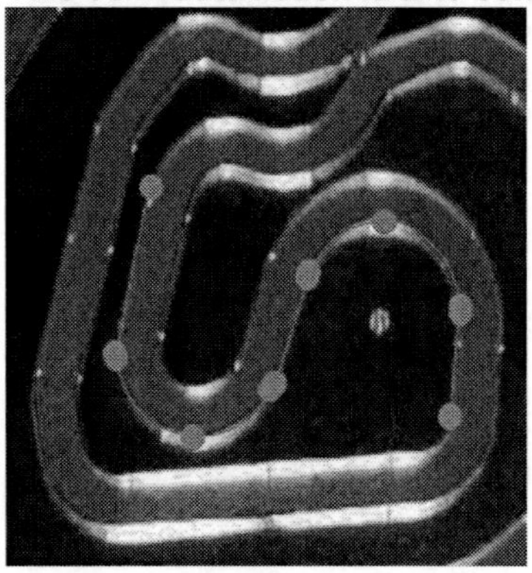

Manual teaching for semiautomated welding

7. (Coin, stretched line).

However, the big innovation came from introducing the information on the weld joint by back scattered electrons. In order to improve precision and speed of teaching an automated scan, perpendicular to the joint, follows the weld line - similar to a race car simulator on a game boy - and stores the weld geometry in the computer. As the unwrapping is a sine shape the center is offset. The computer centers automatically and the line is straight.

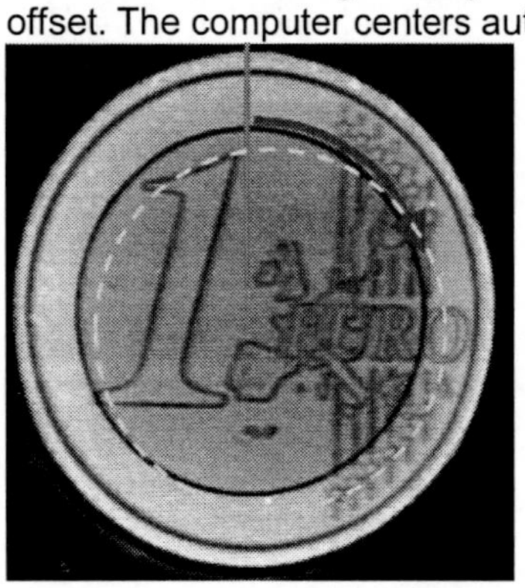

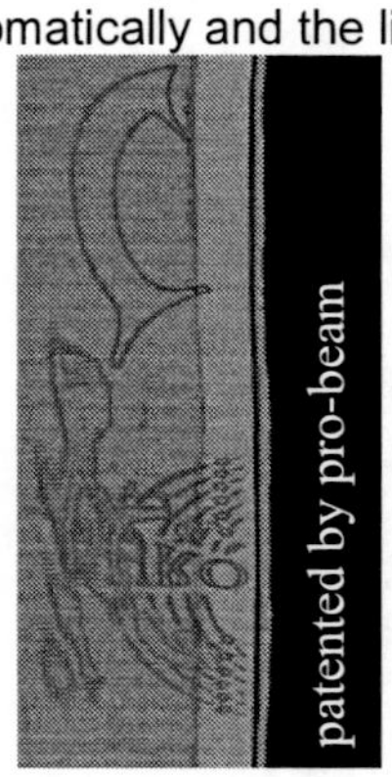

Circular scanning to center a part. The unwrapping is a straight line when centered.

8. (Heat exchanger 2 with bottom line).

This was the big step in technology for automation in EB-welding. Contrary to the previous heat exchanger this contour of the weld joint is not taught manually, but scanned automated perpendicular to the line and stored in the computer.

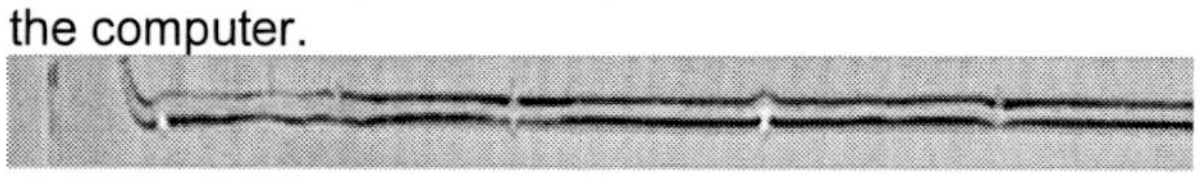

The unwrapping of the scan must result in a straight line. The irregularities result from the secure punch marks.

9. (Valve head pallet).
Precise tools are no longer needed. A loose fit considerably eases loading the pallets and even expanding by warm up of the system is no problem, as the measuring beam is at the same time the welding tool. As the weld is deep and very slender positioning has to be very precise, in the range of 20 micrometers.

Multiple tool with loose fit to ease loading for welding valve head flaps. As the weld has to be very narrow, high precision in position is required.

Valve head flap and cross section of slender weld.

10. (Invisible joint).
On this part there is no center and the weld joint is invisible from vertical down due to the high hub. Still the automatic centering can be performed. The complete surface is scanned in order to produce a full picture. The computer calculates the center for proper welding.

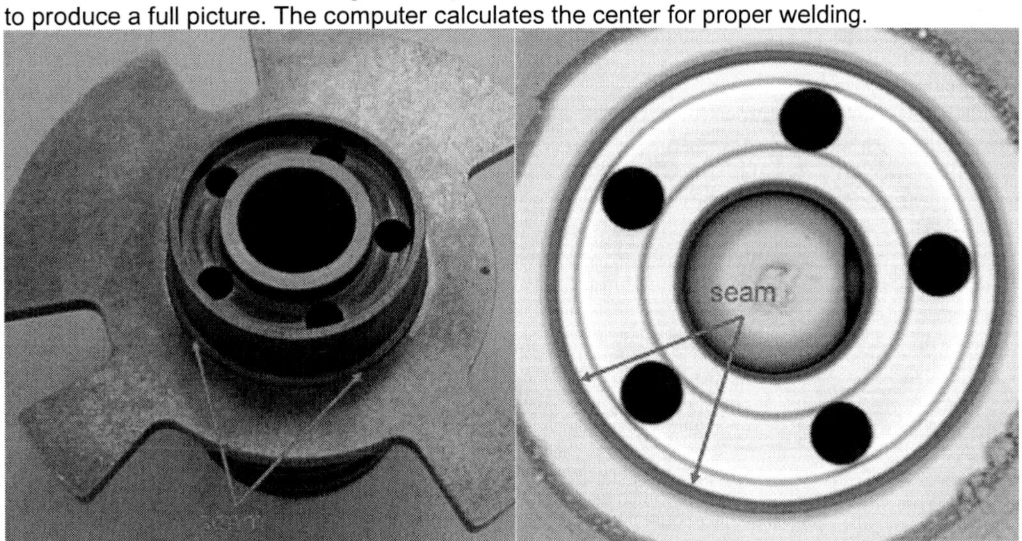

The complete surface is scanned and the computer centers the hub weld.

11. (Siemens, high current connectors).
On these high current switches there are several welds. The contact pieces made from sintered silver-nickel-tungsten and the 1 to 3 copper cables have to be welded to the levers and then it is assembled to a copper block. The deviations of joint position from nominal on workpieces from industrial mass production are considerable.

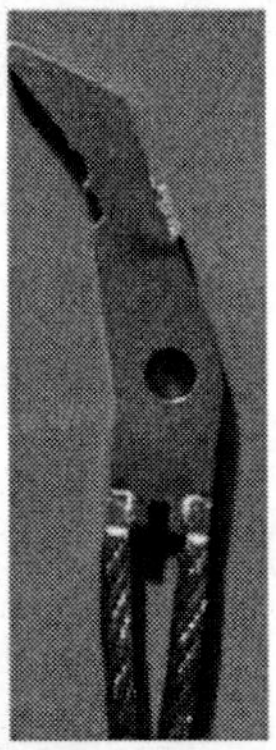

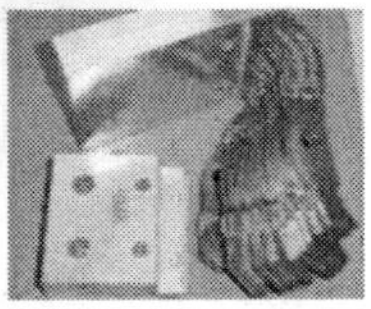
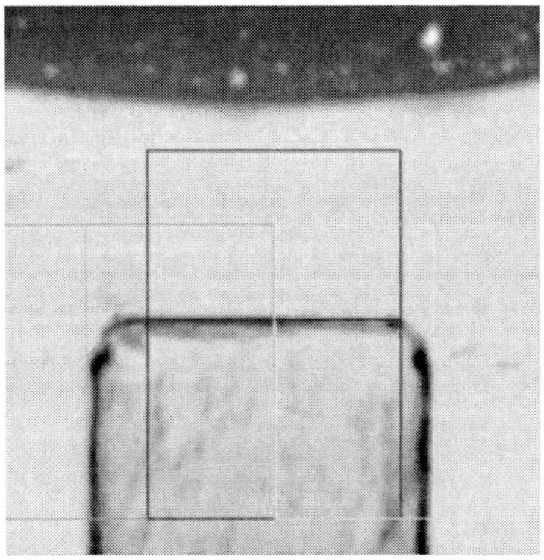

The weld position of 1, 2 or 3 cables have too large tolerances for CNC controlled precise welding without automatic corrections

12. (K6000, Aluplate Ariane, tankbottom).
This is the largest electron beam welder worldwide for subcontract work. The combined gantry and robot motion system has 8 CNC axes and another 8 for the beam control.
In low penetration welds the automated correction of misalignment is done by beam deflection. However when a deep weld has to be performed, the correction by angular deflection would lead to lack of fusion at the root of the weld. Therefore the correction has to be done by the mechanical motion, either by an x-y table or by a robot system or both, securing that the beam hits vertically to the surface.

Ariane fuel tank robot welded from two 70 mm thick aluminum plates, subsequently formed and machined

13. (Bosch sensor).
There are numerous applications where a quality control is required before welding in order to avoid welding of misassembled or misaligned parts and thus producing scrap. The back scattered electrons signal to the CNC the unwrapping of the rotating part and decides whether the part may be welded or be rejected before welding. This pressure sensor tumbles due to bad assembly.

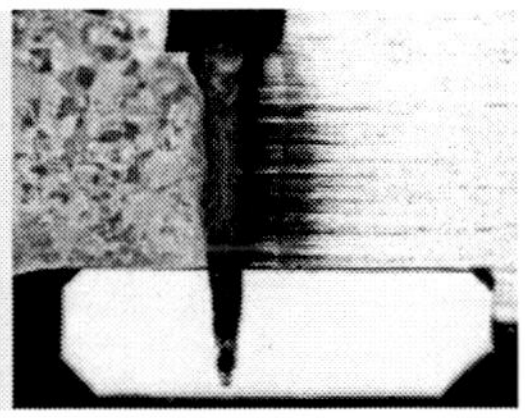
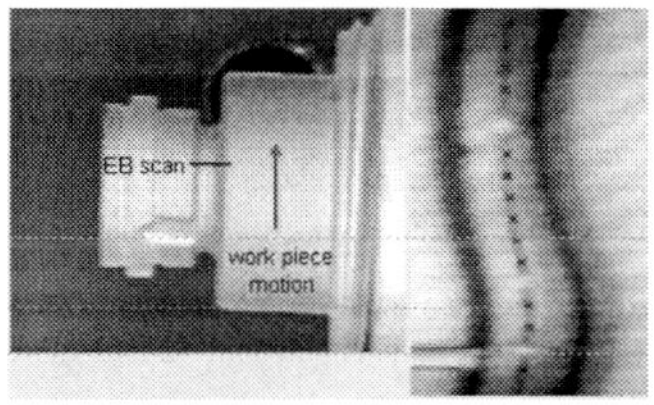

Left: Pressure sensor for common rail injection system.
Right: Weld 0.3 mm wide, 2.3 mm deep. 3 dissimilar materials.

Left: Pressure sensor monitored by backscattered electrons.
Right: The unwrapping of a misaligned circumference joint. A good assembly shows dots in a straight line with no deviations.

14. (Piston rod).
On this piston rod the scan of the unwrapping delivers a straight line and therefore it may be welded. As a good part it is marked with a quality control number.

Electron beam welded piston rod with quality control number.

15. (Gear ring).
Today one piece flow is preferred for mass production. A typical part is this gear ring made from flat material with a dovetail shape of the joint. Parts may arrive at the welding station in varying positions or even upside down. The two parts at the right are inverted. It is no longer necessary to rotate the part into a uniform location; the beam scans the actual position and welds accordingly. The deviations from nominal in both directions are documented.

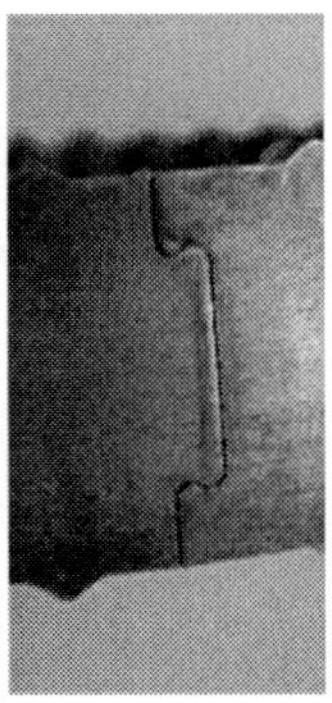
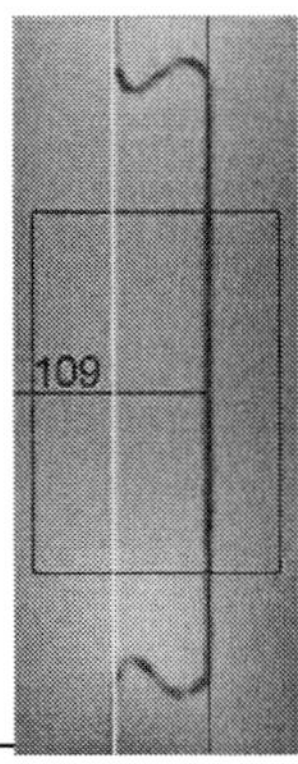

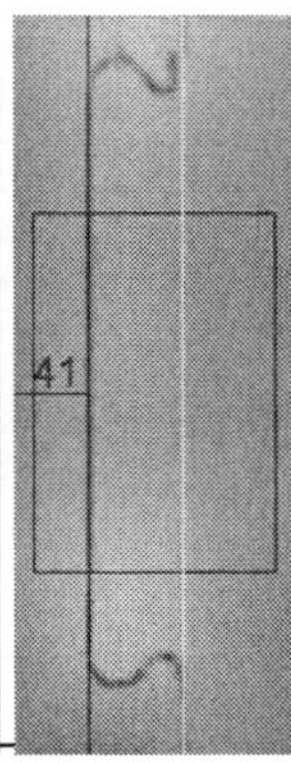

The gear rings may be in different positions or even upside down, the weld will be automatically placed in correct location.

16. (Tar valve).
Heavy parts of martensitic steel use to have strong magnetic fields which are quite difficult to remove. A scan with the electron beam shows where strong deflection occurs and the computer automatically compensates by beam deflection within acceptable limits.

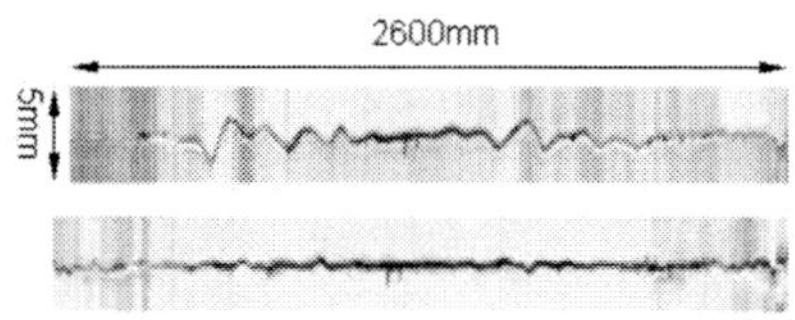

The beam deflection by residual magnetic fields may be smoothed by computer corrections.

17. (Vane ring Turbomeca).
Another category of work pieces carry a large number of joints, e.g. turbine vane rings. A weld sequence plan has to be evaluated for symmetric heat input, taking shrinkage into account. For plausibility a control is introduced by scanning three times along the straight joint and verify that the signals are in line. Including the tack welds this manless operation totals to 1284 scans and 535 welds, performed within 45 minutes, a time, which just allows a diligent operator to assemble the next ring of vanes.

Completed stator assembly with flanges attached.

Detail of vane assembly as welded.

Assembling and welding jig with stator ring in place.

18. (Rolls Royce).
This ring of vanes, the so called front bearing housing, is made of titanium. The challenge for automation is to detect the joint with sufficient contrast and sharpness between the crooked profiles of the vanes. Due to poor accessibility it has to be done at a long distance and oblique. With manual EB welding two operators need 8 hours to finish 1 ring of vanes. The manless operation takes 4 hours and produces much fewer defects. For the complete assembly 19 different welding operations are necessary.

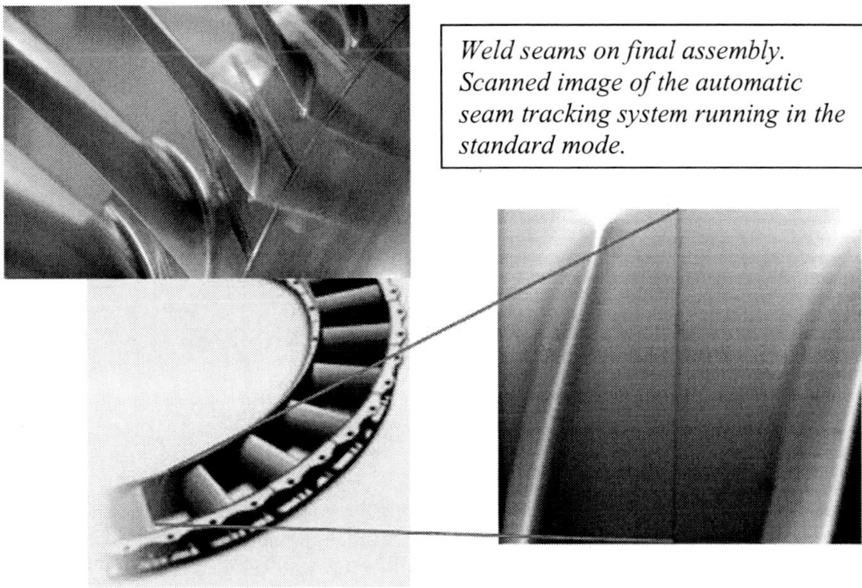

Weld seams on final assembly. Scanned image of the automatic seam tracking system running in the standard mode.

19. (Video clip gear, 3 beams).
The so called multi-beam technology uses actually only one beam which jumps so fast between several positions that the inertia of our eyes or of heat flow sees several spots of impact of this one beam. The temperature of the melt pool stays constant when the beam comes back in sufficient short time. With three weld spots symmetric heat input and symmetric shrinkage is performed, thus preventing offset of the axis of gear and synchronizing ring. The result is a more silent gearbox.

patented by pro-beam

One beam jumps at high frequency between three weld spots.

20. (Catalytic converter).
Another application on a catalytic converter requires very slow welding due to evaporation of coating on the large wire mesh surface. The 4 layers of each pocket are molten together with a butt weld. The electron beam jumps 1000 times per second to each of the 60 welds. By this method the complete catalytic exhaust converter is welded in one run.

Simultaneous welding of 60 filter pockets made of coated wire mesh.

Cross section through the four layer butt weld of each filter pocket.

21. (Flash 8 circles).
An even faster process is called "flash technique". The complete contour of a joint or even several joints is welded within one instant. The time of impact depends on metallurgical and geometrical requirements. The weld contours are programmed and the beam follows those several 100 or 1000 times per second.

The 8 horseshoe like welds of a turbocharger are welded within 1 flash.

Depending on the kind of material e.g. pressure cast aluminium, it may be necessary to preheat, to clean or to degas the part, to weld it and to make a cosmetic seam on the upper weld bead. The manual process does these one after another, whereas a next step of automation allows welding all process steps in one run.

22. (3 beams in sequence, 3 parallel beams).
Each of the three beams in sequence may have different parameters in shape, focus position and power. There are different oscillation patterns, depending on the task.
Of course this can also be performed with three beams parallel, but when the curve starts, the outer beam has to be faster and the inner beam slower than the center beam. As a consequence the beam power has to be adapted automatically according to each weld speed.

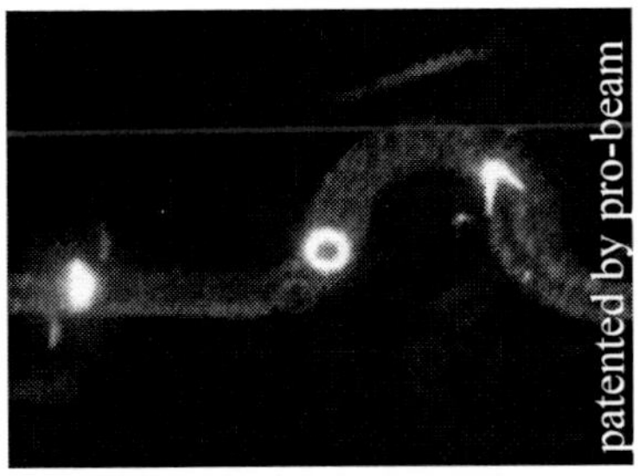

Each beam may have different oscillation and power.

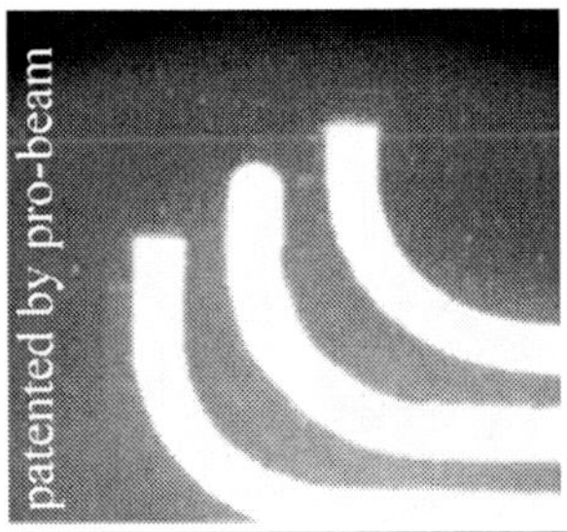

The middle spot follows the track with constant speed, the inner and outer beam have to vary speed and adept power.

23. (Last mission).
Now I would like to present to you the ultimate challenge of automated welding. In order to do this, we have to fly out into the orbit. It is a clear advantage to use electron beam welding in space on two reasons: there is the best vacuum already provided for free and the electron beam offers the best efficiency for energy consumption of all welding processes.

24. (2 astronauts).
My first contact to electron beam technology started in 1969 with an article in the magazine "Der Spiegel" about EB-welding in space during the Sojus mission. These two pictures show astronauts with handhold electron beam guns from the Apollo and the Sojus project. A press campaign started: who was first? For me this campaign was the beginning of a life-long activity with electron beams.

Press campaign in 1969 in "Der Spiegel": who was first? Sojus (left) or Apollo (right)?

25. (Canadarm).
It is by accident that the subject of electron beam welding in space has now become the end of my career. The idea is to assemble structures of a space station by welding with control from the earth. The weld gun may be handled by one of the several robotic arms available on a space station. The capability of an electron beam to see the weld position and transmit the information to a station on ground allows automated welding by remote control from the earth. We still work on this job, but as a proverb says: "not possible is not permitted". Researchers should be motivated by passion. Being successful you feel, despite of all strain, like an astronaut out in the orbit and free from the influence of gravity.

A successful researcher feels free like an astronaut out in the orbit without the influence of gravity.

Ladies and gentlemen,
at the end of my professional career as an electron beam developer and entrepreneur, I would like to take the opportunity to thank the international welding community for their ongoing interest in the fascinating technology of electron beams and their applications. Especially I want to express my gratitude to my closest colleagues in research and development, Dr. Thorsten Löwer and Mr. Michael Hofner with their collaborators. They have done all the work which I have had the pleasure to present to you. For me and my life I can say: "Vision and Mission completed".

Dr. Dietrich Freiherr von Dobeneck

Implementing new technologies in education and training in welding – A precondition to deal with the future in welding

S. Keitel, Halle (Saale)/D, C. Ahrens, Duisburg/D, Ö. Akçam, Ankara/TR, and Y. Xie, Harbin/CN

1 The significance of training and continuing education

Training and qualification are considered by our society to be the most important requirements for recognition and moving up the social ladder. In the industrialised nations as well as emerging or developing nations, they are thus important elements for the development of individuals and, at the same time, a prerequisite for social peace.

Training and continuing education represent powers of production, since demand for qualified personnel is continuously increasing to help realize the creation of constructions and safeguarding of production. The increase in complexity in products and processes can only be accomplished by well-trained personnel.

This is the case for both skilled workers and engineers. It is therefore essential to expand further development of training methodology to its full potential.
One of the most important driving forces for industrialised societies in the last 20 years is the advent of computer applications and the internet. For this reason welding technology and, in particular, training programmes for future qualified welders must meet this challenge head on by developing those advantages that can be gained through the deployment of information technology.

In light of the construction last year (2012) of a new purpose-built training facility for welding technology by DVS, The German Association for Welding and Related Technologies, new computer-supported training methods are being combined with more traditional ones to generate a concept that, above and beyond the field of welding, envelopes the entire sector of metalworking. Figure 1 and Figure 2.

Figure 1 and Figure 2: New development of the BZ RR (Bildungszentrum Rhein-Ruhr) Gelsenkirchen-Schalke Site.

2 The use of computers in direct relation to the training of professional welding personnel

In reference to practical welding training or the training of machine operators, we have identified the following factors whereby computers have changed the world of welding:

- Selection and regulation of welding parameters within the electric power source
- Monitoring of the welder by means of process control (parameter surveillance and surveillance of the welding workshop)
- Ergonomics in the design of equipment
- Body posture for welding – technical evaluation
- Training for the welder – development of hand coordination
- Distance Learning using computer based technologies and internet for theoretical education

Emphasis should be placed on the following elements:

3 Practical training for welders

Impact of the electric power source upon the theoretical knowledge of the welder

It is impossible to imagine our modern power source technology and component handling without computers. Power sources today are capable of storing complete welding technologies in finished programmes, and provide every welder with efficient handling and error-free selection of the appropriate welding parameters.

More specifically, power sources are now capable of, regarding their regulation mechanisms, analyzing the welding process and adjusting to parameters. Particularly through the advent of pulse technology, the drop transfer is controlled, which in the end provides the welder with more time to concentrate on his or her manual skills.

The power source thus assumes tasks that were heretofore dependent upon the manual skills of the individual welder.

The simplified operation of power sources should, however, not lead to the welder being occupied merely with his or her manual skills. It is rather necessary today to put more emphasis on the welder's theoretical knowledge. He or she is the one who has to be able to recognize disruptions in the process and/or an improper selection of parameters. He or she is only capable of doing this when he or she has a thorough understanding of the welding process.

Welders today should also have a basic understanding of control technology, due to the fact that they sometimes have to be able to master both automatic and semi-automatic control processes.

New methods in welding training

A new paradigm has taken place recently in the organisation of welding technology training; this is due to the ever-increasing investment in computer-supported training workstations. This equipment – VWTS or Virtual Welding Training Systems – supports the conventional work of the welding instructor, in that it continuously monitors and corrects the welding student in his or her movements from beginning to end.

In principle, at least 2 methodologies can be implemented today. These are:

Virtual systems
Semi-virtual systems

Figure 3 and Figure 4 illustrate examples for application of the system in welding training. Figure 3 shows a system that generates a virtual weld in the monitor from the movements made by the welder. This and comparable systems guide the learning welder to real challenges through various levels of difficulty. Depending on the type of welding equipment, the workflow movements are however more or less different from the actual welding movements. But they are still quite suitable for the acquisition of fine-motor skills in welding.

Figure 3: VWTS as a virtual system.

As opposed to this, the semi-virtual systems function with a low-capacity electric arc. No real weld is produced, but the welder's movement on the metal plate (which is necessary to start the electric arc) can be followed as a melting-seam line. The welder's movement workflow is recorded during the exercise and he or she receives visual and/or spoken instructions for correction. After welding, the following parameters are made available either in the monitor or printed out: torch position, speed of the weld, torch distance to the surface and an overall comprehensive statement for evaluation.

The advantage of these facilities is that the original welding equipment may be used for the exercises. This involves for example an original torch with a cable-hose assembly with real materials and dimensions. The electric arc also supplies a weak current corresponding to a real glare shield in the form of a welder's helmet. Figure 4 shows such a system as an example.

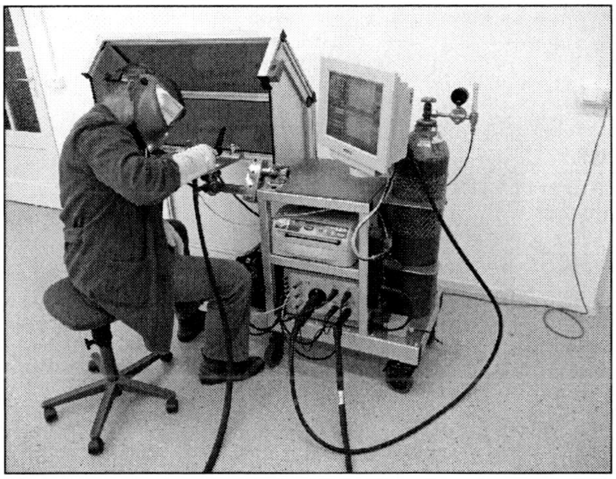

Figure 4: VWTS with real low-capacity electric arc.

Manual skills in welding with an electric arc

Welding is then finally learned using the conventional process, since the energy impact of the electric arc and the dynamic of the welding pool are difficult to simulate. This means that the VWTS today can only cover 20 - 30 % of the overall learning curve for welding.

However, in comparison to a competitive sport, a combination of methods will be used in the future that increases the challenges step by step, thus facilitating the welder's learning curve depending on his or her individual capacities.

In order to support the training process, differentiated VWT systems can be used, for example, in combination or in sequence. It has been found that a suitable method is to combine the purely virtual systems with semi-virtual systems.

At the end of the process, however, the priority is the implementation of learned skills into the welding process, and the application of this into various positions and metal plate thicknesses. This learning method has stood the test of time according to IIW Guideline 089-12.

The typical welding booth for training is illustrated here in Figure 5.

Figure 5: Welder in training.

Use of welding booths and welding cabinets

Training for welders usually takes place in a welding booth to protect those who are not involved from emissions of radiation and smoke. Also, the welding teacher wears a dark helmet during welding training sessions. Training for the technique of welding to date therefore takes place in a kind of "darkroom", in which only the welder's (and the welding instructor's) welding pool can be observed.

VWTS are offering for the first time the possibility of transferring the training process into custom-built, workshop-like training rooms and welding booths.

The welder-trainee can thus be observed and corrected regarding his or her sustainability of movement. Demonstrations given by the welding instructor, or exercises executed by the welding student can thus also be observed and assessed in groups.

The role of the welding instructor

Computer-assisted methods in welding training with the benefit of the VWTS enrich the educational process; they can complement, but not replace, the traditional welding instructor. It is primarily human qualities and handicraft experience that provide young people in training with essential added-value. The welding instructor demonstrates not only theoretical skills and the fundamental workflow movements for welding; rather, he or she makes decisions regarding which methods and which sequences should be used to best facilitate the advancement of knowledge.

His or her communication skills are able to instil in young people the enthusiasm for welding. This is why training sessions with welding instructors are so significant. They are the multiplicators in the training process and determine success. This is true for continuing education within a company, initial training in an educational facility and, as is the case in Germany, a dual system, Figure 6.

Figure 6: Expert assessment of welding results immediately after welding.

Combination with metalworking

If one seeks to thoroughly develop the skills of young people in fundamental training, a complex study of metalworking with the emphasis on welding has proved itself over the years. A good example of this is the workshop of the GSI training centres Rhein-Ruhr which involves training in welding with the application of new VWTS methods combined with training in metalworking. A systems mechanic who has been trained in this manner receives, after exposure to such a dual concept, a nationally-recognized skilled-worker certification.

Figure 7 and Figure 8 illustrate various stations within the training workshop.

The fact that training can be carried out not only in a stationary workshop, but in a mobile workshop as well is illustrated in the mobile training station in Figure 9.

This is also a highly-widespread and modern method for the training of welders. The welding instructor must be more flexible by adjusting to operational requirements, while often times having to overcome language barriers as well.

Figure 7:

Figure 8: Training workshop of BZ RR, training facility of GSI and DVS.

Figure 9: Mobile training station for the deployment of welding instructors in practical application.

4 Education for welding engineers

Welding engineers are considered to be the central focal point for all questions regarding quality control in welding processes. Their technological qualifications therefore determine the quality of welding constructions and products. This is why the qualification of expert welding engineers, within the context of welding-supervisor training, carries the highest priority.

In Germany in recent years, approximately 1,000 IIW welding engineers per year have been trained. China grows up on the same level. New markets like Turkey become on the same way. The qualification of welding design engineers has also developed quite positively, since they represent the multiplicators for new designs and applications for modern welding processes in the creation of products.

The successful increase in qualifications can be attributed to, in particular, the deployment of new learning methods, since approximately 350 participants are trained by GSI per year via CBT – Computer Based Learning or the so-called Blended Learning. An overview of learning content, which is offered today through Distance Learning, is illustrated in Figure 10.

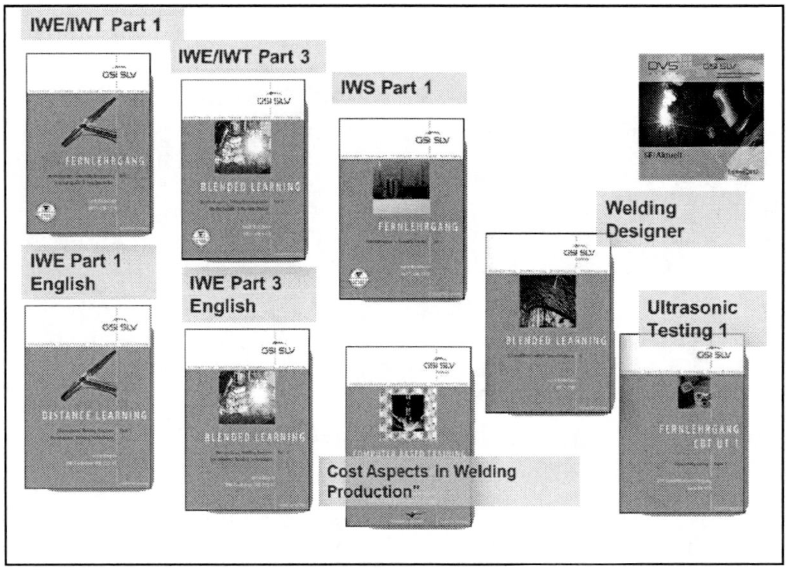

Figure 10: Distance learning courses of GSI, enterprise of the German Welding Society.

Distance learning courses require from the participants a high level of self-motivation. Independent mastering of the learning material is facilitated with visual support, as presented in Figure 11.

The course variation "Blended Learning", which has been specially implemented for the training of welding engineers, is based on 50 % self-study and 50 % attendance in the classroom. The latter offers participants the opportunity of posing up-to-date issues and questions directly to the instructors and mentors, while providing the teachers an environment for immediately integrating latest information (e.g. technical standards) into the course.

The reason that Distance Learning courses are increasing in significance can be attributed to the opportunity to achieve a qualification through an extra-occupational program, thus corresponding to the trend of life-long learning.

In this context, the impact of the internet must be respected, since questions from course participants are directly forwarded to the mentors and instructors. Modern terminal equipment and broad-band data connections support this trend as well, Figure 12.

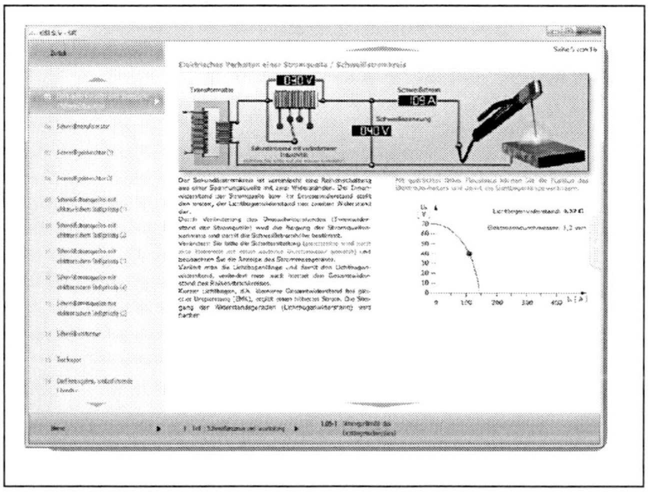

Figure 11: Screen with animations – example taken from a Distance Learning course.

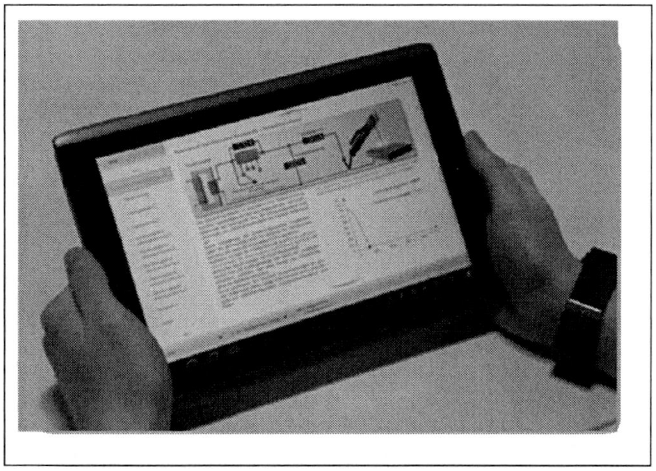

Figure 12: Modern Tablet-PC as a terminal device for continuing education in welding technology.

5 Summary

Based on examples taken from practical training of welders and continuing education for welding engineers, it has been illustrated here how the methodologies for transfer of knowledge have been enhanced for welding technology as well through the influence of new technologies and with aid from computers and the internet.

This was necessary in this respect in order to develop, with the aid of new technologies, supplemental effects such as:

- More flexible training,
- Individualisation of the training,
- Structured learning

On the other hand, it is also important to enable the preservation of resources, e.g. in the form of travel time, energy, less paper, etc.

This has a long-term impact upon the overall training process and must be taken into consideration today in planning and investment for new educational facilities. Moreover, computer-based training methodologies form the basis for more-robust internationalisation of welding technology, as can be seen from the interest of many countries worldwide. It is no surprise that emerging markets join these new trends faster than the old economies.

Sustainability through innovative welding processes in automated applications

A. Burt, C. Fink, Mündersbach/D, and P. Kuebler*, Rocklea/AU

* The author has written the chapter 5

Sustainable action is the conscious and careful handling of available resources. Sustainability in welding technology requires not only the saving of energy, raw materials and emissions but also of welding time and, with it, costs. These savings can be achieved using highly dynamic, energy-saving power sources and innovative welding processes. In an industrial concern, quality and economic efficiency have top priority. To enable sustainability, economic efficiency and quality to complement each other in the best possible way, a precise knowledge of all interrelationships and the interaction of the different components is a fundamental requirement. Using the example of an automated surface welding process on an actual welding task, these three interrelationships and their relevance from an overall perspective are clarified below.

1 Surface coating process

The productivity of machines and plants is determined to a large degree by the properties they exhibit in dealing with the demands which arise. For certain applications, the corrosion and wear properties or their tribological behaviour are the significant factors in designing components. Special surface-coating processes have been developed to improve the surface properties of components. These techniques permit the economic and resource-efficient use of materials. In the field of thermal coating processes, a general distinction is made between the surfacing processes of cladding, hardfacing and buffer-layer welding. Cladding is primarily used for corrosion protection of the parent metal, whereas hardfacing is used for wear protection. Buffer layers are applied to the parent metal as intermediate passes for dissimilar surface coatings (1) (2).

In thermal surfacing, a wear- or corrosion-resistant surface material is applied to a dissimilar, often unalloyed steel using arc processes, in order to improve the properties of the surface in line with respective demands. The firmly bonded connection between a sufficiently elastic component and functional surface and the associated high thermal conductivity, independence from component geometry and creation of dense layers offer the essential advantages of surfacing as compared with thermal spraying. The properties of the functional coating depend very much on the dilution, i.e. the ratio of fused parent metal to the entire fused volume. The degree of dilution of the functional layer can be determined by means of cross-section polishes or via the ratio of the chemical analysis of the layer areas. Depending on the welding task, material, surface layer and degree of automation, dilutions of >5–30% can be achieved at deposition rates of up to 40 kg/h (3). In the TIG surfacing shown, in practice the dilutions are >10% at deposition rates of up to 4 kg/h.

2 Alternative TIG-DC process with AC hotwire

The TIG-DC with AC hotwire welding process is a newly developed alternative process in which the arc is produced between a non-consumable tungsten electrode and the workpiece by direct current. The filler material is fed mechanically and preheated by the application of an AC current ranging between 30 A–170 A at voltages ranging between approximately 2 V–8 V (output between 60 W and 1.4 kW). The combination offers the advantage of achieving a continuous and reproducible process with a considerably higher deposition rate in comparison with a manually fed welding rod. Here, the fed welding wire electrode is heated with alternating current instead of direct current. Electrical engineering has shown that a magnetic field arises around energised wires and that its direction of action changes depending on the direction of energy flow within them.

The magnetic field formed by hotwire with alternating current flowing through it and the magnetic field formed by the energised DC arc interact with one another by reciprocal attraction and repulsion of the arc column. For this reason, the arc can easily be deflected by a magnetic field due to the highly flexible plasma column.

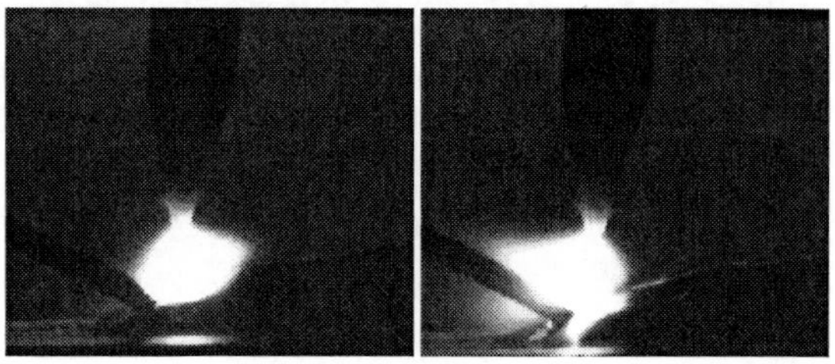

Fig. 1: Weaving of the arc due to the alternating magnetic field in the welding rod

Normally, deflection of the plasma column is undesirable for arc processes. These properties make TIG-DC with the AC hotwire process advantageous. By applying alternating current to the filler material, a constantly changing magnetic field is created, as the magnetic field of an energised wire reverses depending on the direction of the current. Due to this effect, the arc is constantly deflected at the frequency of the alternating current (attracted or repulsed, Fig. 1).

The result is therefore high arc currents on the electrode and a high deposition rate due to the preheated filler material. Due to the arc which can be deflected at the adjustable alternating current frequency, low penetration and low dilution is achieved with the parent metal. In this way, deposition rates can be attained similar to pulsed GMAW, but without the undesirable high dilutions.

3 Trials for surfacing as a corrosion protection measure

Unlimited corrosion protection exists in surfacings in which at least 12–13% chromium can be demonstrated in all areas of the surfacing. For many applications, the limits are considerably narrower. Accordingly, the lowest possible degree of dilution with the parent metal must be the goal when welding components exposed to gases containing chloride. The absence of cracks in the functional layer must also be guaranteed. The aim of the test was to attain the limits of the newly developed welding process in terms of boundary parameters and to achieve the lowest possible dilution whilst at the same time maintaining a high deposition rate. By means of preliminary tests, the test rig was decided upon in respect of the inlet direction and the gap between the filler material and the electrode. The gap between the electrode and the workpiece should vary between 4 and 6mm and the gap between the hotwire and the electrode should be 5 mm at an angle of 20–30° to the workpiece (Fig. 2).

The frequency of the adjustable hotwire alternating current, which from an equipment perspective is infinitely selectable between 50 and 200 Hz, was fixed at 100 Hz. The balance of the hotwire alternating current, i.e. the proportion of positive and negative half-wave for each frequency, was able to be varied within limits of +/-30%. Alteration of the waveform of the alternating current between the sinusoidal, rectangular and trapezoidal forms is also possible. In principle, the sinusoidal waveform offers the advantage that the highest currents fit closely in the maximum of the period. The higher the current, the greater the magnetic field, the more pronounced the deflection of the arc in the maximum current. Viewed over the entire period, the mean value of the current is identical for all waveforms, only the maximum current varies.

In TIG welding, inert gases or inert mixed gases are always used as shielding gases. In order to assess the properties and effects of the shielding gases on the flow and wetting behaviour, the surface tension of the molten metal and the stability of the arc, in accordance with ISO 14175 work was carried out using the protective gases I1-Ar, I3-ArHe-30, I3-ArHe-50 and R1-ArH-5. The trials using I3-ArHe-50 delivered the best results.

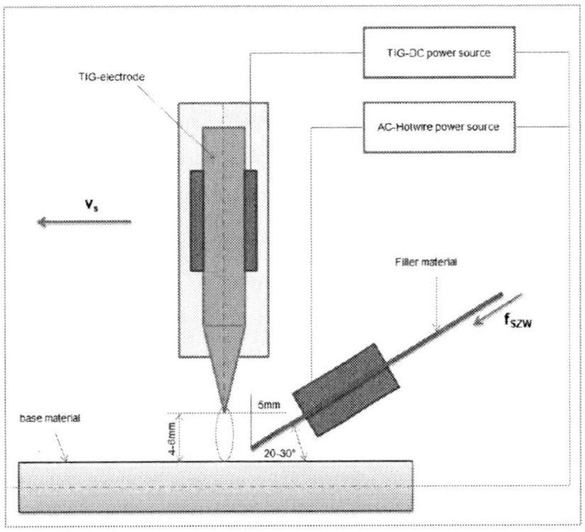

Fig. 2: Schematic diagram of the TIG-DC test rig with AC hotwire

For applying the corrosion-resistant layer onto the unalloyed S355J2 (1.0577) structural steel, an S Ni 6625 nickel-based solid wire electrode in accordance with ISO 18274 (2.4831) in Ø1.2mm was used. This contains, as the main alloying elements, 22% chromium, 9% molybdenum, 3.6% niobium and the rest nickel.

In the trials, melt runs were applied to the plate surface. The trials were carried out fully mechanically on a 5-axis robot with the EWM Tetrix 551 Synergic as a power source, with an additional EWM-AC hotwire power source in the form of the Tetrix 200 AC. A WCe-doped electrode in Ø4.0 mm was used as the TIG electrode. The analysis was carried out allowing for the most minimal dilution at the same time as the maximum deposition rate.

The boundary parameters were determined in the initial trials. What is interesting is, that the deposition rate, where welding parameters are otherwise identical, increases by +48% following a variation in the balance on the AC hotwire feed from +30% to -30%. This is equivalent to an increase in wire feeding from 8 m/min to 12 m/min and a rise in the deposition rate from 4.3 kg/h to 6.4 kg/h, while penetration remains lower at the same time.

It is apparent from Fig. 3 that with sinusoidal alternating current at a frequency of 100 Hz, a balance of -30%, an arc current of 370–380 A and an AC hotwire current of 130 A, reproducible and usable results can be expected. Trials 2 and 6 show that the dilution is under 8% and 7.5% respectively, at a deposition rate of over 6 kg/h (12 m/min wire feeding) and at welding speeds of 1.0 m/min (Fig. 6). At identical deposition rates, this results in an approximately 70% lower dilution with the parent metal compared with pulsed GMAW surfacing. In addition, no lack of fusion or cracks whatsoever could be detected in the tests. However, all the trials highlighted give rise to a reproducible and stable welding process.

	1	2	3	4	5	6	7
Current [A]	330	370	370	380	380	370	380
Welding speed [m/min]	1,00	1,00	1,00	0,90	0,95	0,95	0,95
Filler wire speed [m/min]	9,0	12,0	12,0	12,0	8,0	12,0	9,0
Hotwire current [A]	90	130	130	100	130	130	110
Wave form	sine wave	sine wave	sine wave	sine wave	sine wave	sine wave	sine wave
Frequency [Hz]	100	100	100	100	100	100	100
Balance [%]	-30	-30	-30	-30	30	-30	30
Hotwire feed direction	F	F	S	S	S	F	F
	trailing torch		90° to direction of travel			trailing torch	
Speciem number	E14	E15	E16	E17	E18	E19	E20
Deposition rate [kg/h]	4,82	6,43	6,43	6,43	4,29	6,43	4,82
Dilution [%]	>15	<8	>15	>15	>15	<7,5	<7,5

Fig. 3: Summary of analysis results

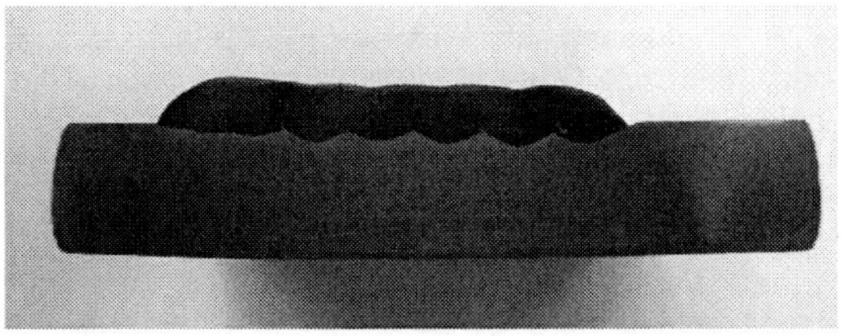

Fig. 4: Balance +30%; wire feeding 8m/min

Fig. 5: Balance -30%; wire feeding 12m/min

Fig. 6: Clad layer in the cross-section polish

In order to calculate the dilution in terms of area, the test pieces were measured in the cross-section polish with the aid of a computer (Fig. 7).

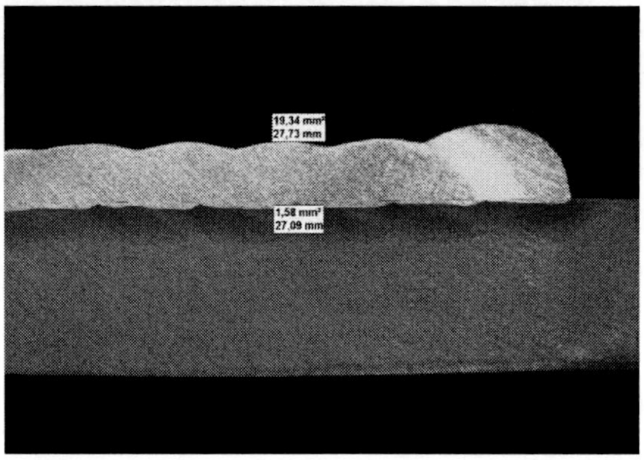

Fig. 7: Test piece E19, cross-section polish with calculation of dilution

In further trials, it was possible to determine the effect of the TIG-DC with AC hotwire technique on the distribution of carbides in a hard surfacing operation. For a wear protection, it is important that the carbides are not destroyed during the welding process and are present in fine distribution in the surfacing following treatment. Only in this way can a wear-resistant and durable surfacing be achieved. Thanks to the new alternative process it is now possible, at deposition rates of up to 6 kg/h, to achieve an abrasive-resistant application with the most finely distributed carbides. In the trial, a tungsten carbide flux-cored wire was applied on an unalloyed structural steel in the identical test rig.

In practice, this welding process is being accepted on the strength of the positive trials. One international pipe manufacturer is thus producing abrasive- and chloride-resistant functional layers on the inside of large-diameter, unalloyed structural steel pipes on an automated basis. In the example, 12 welding torches are attached to a lance with filler wire and apply the functional layer to the individual pipe segments (length up to 12m). Thanks to the new alternative TIG-DC with AC hotwire process, the manufacturer has been able to shorten his cycle times measurably by 30% whilst increasing the quality and durability of the functional layer.

4 Outlook

At a time of ever-increasing demand for raw materials and primary energy, sustainable and resource-conserving action will gain considerably in importance. Many business enterprises are already taking the duty of handling raw materials and energies and the conservation of the environment and humankind very seriously. By virtue of its savings in overall energy demand due to shorter manufacturing times, a reduction in non-productive times due to "non value-adding activities" and a significantly improved welding task outcome, this new welding process is also contributing to a cost-conscious, sustainable and resource-conserving overall process.

5 Examples of sustainable innovative welding and cutting processes in automated applications in Australia

P. Kuebler, Rocklea/AU

Robotic welding is not new in Australia. It has been used in the automotive and manufacturing industries since the 1970's. Although Australia has not been a leader in the development of robotics, it has produced some innovative applications which are world leading. This paper describes a number of these welding and plasma cutting applications.

Robot welding of large fabrications took a major step forward when a robot system was developed by Robot Technologies Australia (RTA) for Thales to weld out the hull of the Bushmaster armoured personnel carriers for the Australian defence forces. The Bushmaster hull weighs 7 tonnes, is 7m long and involves about 600m of welding, in many cases multi-pass. Hull sides have many penetrations and welded attachments which distort the high strength steel plate. The hull is assembled and tack welded, then placed in a positioner which can be manipulated 360°.

To gain access to the internal welds, the robot cell uses standard 6kg payload robots mounted on large 300 kg pick and place robot as shown in Fig. 1. These robots run on 10m long tracks (shifters). The key to success of this configuration is the use of laser joint location and tracking. RTA developed algorithms that shift the robot path from the laser location data. Once welding commences, laser tracking adjusts the robot path in real time. Smaller robots are used for preheating the armour steel and monitoring its temperature.

Robot welding of the Bushmasters has increased throughput by 400% and ensured the viability of their manufacture in Australia. For the small family owned RTA business, this was a significant achievement.

BOC began marketing EWM welding equipment in early 2010 and partnered with RTA and other robot integrators to supply turnkey automated welding systems. In 2012, BOC and RTA with our customer Monadelphous developed a concept for automating the cutting, bevelling and welding of piling for a jetty and coal wharf on the east coast of Australia. Monadelphous is a major resource infrastructure construction company. The driver for this investment was the shortage of skilled welders and the rewards for early completion of the project.

A Kawasaki RS 15X long reach robot equipped with a Thermal Dynamics Ultra-Cut 200 plasma cutting system mounted on top of a driven pile cuts and bevels a U profile in the top of the pile which later accepts the headstock beam. The robot uses a laser and touch sensor to map the surface of the pile to identify the pile's shape and orientation. This data is used to re-calculate the initially programmed robot path using algorithms jointly developed by RTA and Monadelphous.

Fig. 1 Robot on robot welding trials of Bushmaster hull

Fig. 2 Robot arrangement in Bushmaster production cell

A typical robot plasma cut and bevelled component is shown in Fig. 3. The cutting and bevelling accuracy is ± 2mm over 10m. One robot can do in one hour what previously took 2 men 4 to 6 hours. This is primarily due to the accuracy of the robot path coupled with the cutting speed and quality of the high definition plasma cutting system.

Other robots equipped with EWM alpha Q 552 multi-process welding systems are mounted in place to weld the headstock to the pile. The team evaluated a number of processes and found that gas shielded rutile flux cored wire provided the greatest reliability, quality and deposition rates of up to 6kg/hr even in the vertical position. The full penetration V butt weld takes 3 to 8 passes depending on the pile wall thickness which varies from 16 to 25mm. Fig. 4 shows a multi-pass vertical up (PF) butt weld in 16mm material.

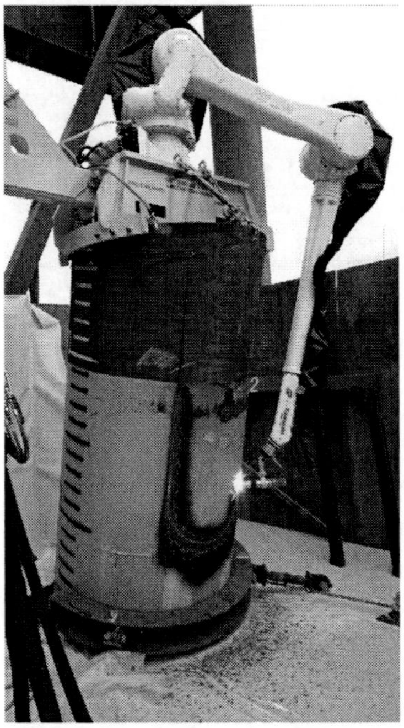

Fig. 3 Robot plasma cutting and bevelling a pile

Fig. 4 Multi-pass vertical and overhead robot welding

In the late 1980s, attempts were made to automate the fabrication and maintenance welding of dragline buckets for the coal mining industry. These buckets have capacities of up to 120 m³ and weigh up to 100 tonnes (see www.cqmsrazer.com/hurricane). A tripod robot was mounted on a forklift to reach the up to 3m high sides. To date, robots are not used for mining bucket fabrication and maintenance although a great deal of development work has been done in the last year.

This work involves the use of gas shielded flux cored wires and multi-pass procedures capable of welding plate up to 100mm thick and castings up to 300mm thick. Jigging by conventional methods is not suitable for large structures. The use of laser seam finding and tracking enables the simple programming of multi-pass applications using algorithms developed by RTA.

Another interesting application is the welding of large diameter cable reels. Australian Cable Reels manufacture reels ranging up to 4.2m outside diameter. Fig. 5 shows a Kawasaki RA 10L robot equipped with EWM Phoenix 352 and Servo Robot SFD Find laser camera. The components are tack welded by hand, the assembly is placed on a carousel and indexed under the robot operating envelope. This methodology is consistent with past practices and requires minimal jigging thanks to laser vision. The welding process used is conventional MIG/MAG.

Fig. 5 Robot MAG welding of 4.2m diameter cable reels using laser seam finding and tracking

Acknowledgement

The author thanks Robot Technologies Australia and Monodelphous for permission to publish this information. We also acknowledge the vision and support of Perry Rae (Quality Superintendent) and Adriano Marin (Senior Project engineer) of Monodelphous.

Literature

[1] Hartung, F. *Verfahren des Auftragschweißens.* [Buchverf.] H.D. Steffens and W. Brandel. Moderne Beschichtungsverfahren. s.l.: DGM-Informationsgesellschaft Verlag, 1992.

[2] Schreiber, F. *Verschleißschutz durch Auftragschweißen: Werkstoffauswahl und Anwendungstechnik.* Willich: Durum-Verschleiss-Schutz GmbH.

[3] DIN EN 1011-5 *Empfehlungen zum Schweißen metallischer Werkstoffe – Teil 5: Schweißen von plattierten Stählen.* Berlin: Deutsches Institut für Normung e.V. [German Institute for Standardisation], 2003. October 2003.

Automated laser manufacturing of hydraulic tanks, switch boxes and control cabinets

N. Höppe, Obernburg/D

Laser materials processing has opened up ever more areas of application in recent years and industrial fabrication is no longer conceivable without it.

While it has, until now, predominantly been a matter of portraying the actual process (e.g. with a robot in a self-contained small unit), interlinked, fully automatic installations have also been the subject of inquiries and put into effect for quite some time.

These installations are utilised for the laser welding of hydraulic tanks or medium-voltage switchgear housings. In both installations, the jigs are transferred into and out of the laser welding cell and the welding is carried out in a fully automatic operation in each case.

With regard to the laser cutting of switchgear cabinets, the advantage lies in the automated offline program creation in order to be able to fabricate them in an economically viable way right down to the quantity of one.

Application of the laser remote welding technology with 3D sensors in ship and civil engineering

C. Emmelmann, J. Wollnack, M. Kirchhoff, F. Beckmann, G. Cerwenka, W. Fricke, C. Robert, Hamburg/D, A. Drenker, Aachen/D, F. Niese, Saarbrücken/D, R. Wagener, Darmstadt/D, M. Kogel-Hollacher, Gaggenau/D, C. Paul, Haiger/D, J. Thieme, T. Westphäling, Burbach/D, T. Witolla, H. Deymann, F. Boekhoff, Hamburg/D, P. Lummerzheim, Kiel/D and Hamburg/D, and N. Worm, Hamburg/D

1. Introduction

In a few years cruise giants will probably call at many new and smaller ports – without danger of running aground. New ideas reduce weight of ships, save fuel and deplete the draft. Modern shipbuilding provides highest demands on joining technology in steel construction and quality requirements of the classification society associated with large dimensions of three-dimensional (3D) components. According to growing competition from East Asian shipyards the need and necessity increase to improve the productivity and the quality of steel construction continually. The joining technologies and the technologies adapted to these methods are most important for the productivity and the quality of the result. In addition the shipbuilding industry contemplates using more robot-aided automation solutions.

Introducing laser welding technology, which exists in the shipbuilding industry in the panel manufacturing since a few years, new constructions of the hulls can be realized. The comparatively cool working method means a big asset for the massive and large volume shipbuilding in contrast to the conventional welding technology. The material does not nearly heat up as much as before. The fine nevertheless high-energy laser beam allows almost distortion-free welding of thin metal sheets, as they are not available with conventional welding processes in a comparable quality. Using the laser remote technology, which means welding with large distances between the workpiece and the working tool of about 1.5 m, previously difficult reachable areas in ship and civil engineering can be welded in all three spatial directions in the highest quality.

2. Motivation and aim

The research project QuInLas (funded by the German Federal Ministry of Economics and Technology based on a decision of the German Bundestag) explores new design approaches in order to manufacture ships faster and design them not that heavy. In particular joint handling of ship structural components is one of the most important manufacturing tasks in shipbuilding. The development of the last years shows new welding processes like the laser beam welding are used increasingly next to the old conventional processes like MIG or MAG welding. Because of the large number of currently available methods, a critical comparison of all procedures is necessary for the selection of specific components during new design of them and its production planning. This includes just the knowledge of the process specific and constructive design opportunities to realize new designs coupled with a consistent lightweight construction next to an assessment of production technological, material and quality specific and economic potentials of the individual techniques.

The aim of the project QuInLas is to develop process and system technology and also valuation methods for the new laser joining technologies like laser remote stitching and laser hybrid welding. They are necessary to fulfill future requests to three-dimensional welded constructions in the shipbuilding. A manufacturing demonstrator will be developed for the three-dimensional flexible laser remote and laser hybrid welding with high automation potential for an extensive and productive use of thin metal sheets and process-induced low thermal distortions of the laser joining for innovative ship constructions. In order to reduce the quality assurance effort, a direct statement about the joining and process quality will be given by an integrated quality diagnostic and valuation system using deposited calculation models. A special post-processing automated ultrasonic inspection examines the joining connection specifically for discontinuities.

3. Project composition and structure

The research group, under the direction of the Institute of Laser and System Technology (iLAS) of the Hamburg University of Technology (TUHH), consists of a total of 11 partners. These include five research institutes in the fields of laser technology (iLAS, Hamburg; ILT, Aachen), the quality diagnostic (IZFP, Saarbrücken) and the fatigue strength (SKF, Hamburg; LBF, Darmstadt, Germany). As well as six industry representatives (Meyer Werft GmbH, Papenburg; ThyssenKrupp Marine Systems Blohm + Voss Shipyard GmbH (BVS), Hamburg; Carl Cloos Schweißtechnik GmbH, Haiger; German-ischer Lloyd AG, Hamburg; IPG Laser GmbH, Burbach; Precitec GmbH & Co. KG, Gaggenau).

The project is divided into three working groups. The first group "Innovative ship design", composed of the institute LBF and SKF, the Meyer Werft and Blohm + Voss, deals with the assessment of novel structures possible due to the application of laser technology as well as the calculation and analysis of the lifetime of the laser-welded structures. The second group "3D laser manufacturing demonstrator" explores the process of the laser remote and the laser GMAW hybrid welding with laser powers up to 30 kW. Moreover an important element of the group assignment is the development, design and construction of the laser manufacturing demonstrator where the demonstrating workpieces will be welded. The third group "Integrated quality assurance", consisting of the institutes ILT and IZFP and the company Precitec GmbH & Co. KG, explores the possibilities of in-process diagnostic and post-process diagnostic with an ultrasonic inspection system.

4. Component design für shipbuilding

4.1. Innovative ship design

To realize the full potential of the innovative manufacturing methods and test its capabilities in realistic conditions a demonstrator section was developed in cooperation with the shipyard industry. The demonstrator includes shipyard specific constructions and the most challenging designs for the welding process that can be found in commercial ship building applications. All the relevant components like knee sheets, seal sheets, walls, etc. are integrated in the demonstrator and laid out with a minimal distance between single components found in shipbuilding. Due to the worst-case-scenario design of the welding demonstrator the research results match the challenges and can be applied to commercial ship sections. Figure 1 shows two representative examples of the innovative design approach taken in the project.

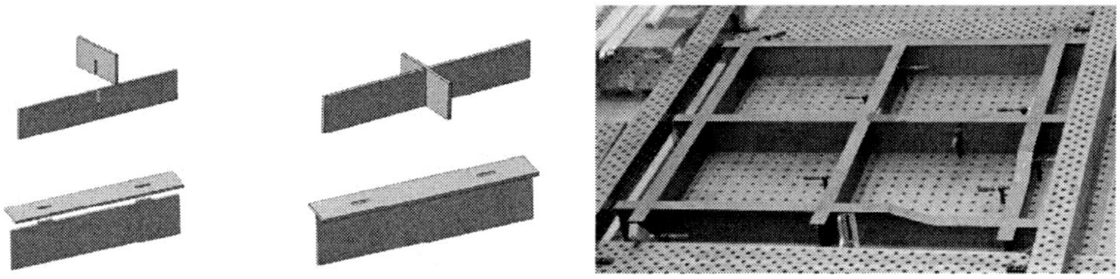

figure 1: left: overview of tested, innovative connections – combed cross joint (top) and strap-stay connection (bottom); right: real setup of the concepts [source: iLAS]

The new design of the comb connection, shown on the top left, introduces a slot into the sheets during the cutting process so that they can plugged into each other in the set up for a fixed positioning. A displacement during the welding process is prevented by this design. The strap-stay connection design on the bottom left of figure 1 pursuits the same principle. Due to flags in the stay and matching notches in the strap both components can be fixed to each other.

5. 3D laser remote manufacturing demonstrator

5.1. Development of laser remote welding for ship structures

Laser remote welding technology is used successfully for several years, particularly in the automotive and automotive supply industry. The idea is the combination of beam moving over a high dynamic level associated with a focal length of about 500 mm. This principle allows laser spot velocities of 60 m/min and more. In particular, the technology provides jump speeds of more than 300 m/min to reduce the idle time between the stitch welds to a minimum with great effect especially in case of many seams.

Because of two reasons in the field of shipbuilding, this technology has not found application yet. On one hand laser beam power > 10 kW, as they are needed for materials used with thickness > 10 mm, was not available in the past. On the other hand, the common practice of manually stitch welds is sufficient for the subsequent full seam welding by the GMAW method. However the stitch welds welded by hand cannot be used with the laser only or the laser GMAW hybrid technology, because the design throat thickness is too large and the seams are dependent in their expression by the welder. Furthermore components obviously utilized in shipbuilding cannot be accurately positioned below one millimeter. As a result of that, the required time increases significantly for the setting before the joining process. Every single seam has to be recorded through many nodes along the welding path for a robot-assisted machining. Even a possible interface to a planning environment system supports the tool setter in preparation of the sequence control of the machine tool and the machining area.

The project QuInLas accesses these prevailing problems that sharply hinder or even prevent further automation with the development of a novel 3D laser remote manufacturing demonstrator, which is specially designed for ship and civil engineering. The new system of the Institute of Laser and System Technology (iLAS) is able to record, measure und correct all path parameters of a 3D welding task automatically by optical sensors combined with an intelligent calibration and image processing as well as path calculating algorithms. Afterwards this system performs the processing task with high precision. Figure 2 shows a prototype of the system demonstrator.

figure 2: operational laser remote manufacturing demonstrator with active green laser measurement analysis in the process of a "component" [source: iLAS]

For the calculation of the path, a green colored laser with less power is used to scan the surface of an object. With the help of a multi-camera system, which is sensitive to the specific wavelength, the optical path length of the reflected light is analyzed and therefore the surface structure and welding seam can be detected. Details of the current studies on a specimen can be seen in the images of figure 3.

figure 3: excerpts of studies on laser remote stitch seam welding on lap joint [source: iLAS]

Based on the data collected on all seams of the component and compared with the CAD data the previously calculated path for the seam can be corrected. Afterwards stitch welds for fixing the sheets are welded according to a predefined scheme, which is designed to minimize the distortion of the welded seams, like figure 4.

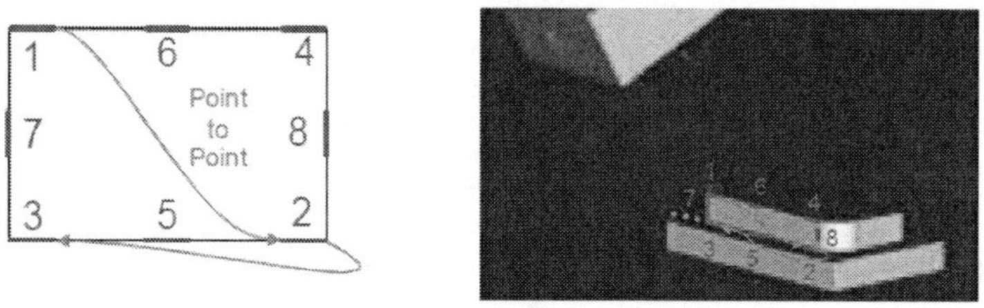

figure 4: one way of a temperature-induced minimized and distortion low stitch seam scheme for laser remote welding [source: Dr.-Ing. habil. J. Wollnack]

The execution of stitch seams leads to substantial independence between the robot and the laser spot movement. The seam and the robot motion are not coupled during stitching. As a result, the system obtains the great freedom in three-dimensional space, especially helpful for the difficult positioning of large structural components used in shipbuilding. Therefore the system is able to significantly reduce the processing time and cost. The remote laser scanner system to be developed is flexible and can be installed on any stable machine system.
Further developments in the field of laser technology have produced beam sources that achieve an optical efficiency of up to 30 % and provide considerable laser power of over 10 kW. However, this is not sufficient for the current requirements in the shipyard. The remote technology requires more laser beam power and higher beam qualities so that IPG Laser GmbH has developed a special laser source for the project QuInLas, which meets the requirements of daily use in the shipyard industry at the same time. This laser is nowadays the most powerful, civilian solid-state laser beam source in the world with a laser beam output of 30 kW. It is possible to split the output, so that the laser remote head and the laser GMAW hybrid head can be provided by laser power in parallel. The process fiber has a diameter between 300 microns and 400 microns. Initial tests of the Ytterbium laser system (YLS), combined with the novel scanner design, show the great potential to revolutionize the shipyard manufacturing and civil engineering in the near future.
For the process, the above-mentioned combed walls are plugged together and the laser remote technology automatically places the stitch welds in a three-dimensional working area of 1000 mm x 1000 mm x 250 mm. The workpiece becomes sufficiently rigid in order to subsequently finish full seam welding with the laser GMAW hybrid technology. Under laboratory conditions and later on under real working conditions the system should be able to join one meter in less than 30 s and is therefore faster and more accurate than the fastest shipyard welders. Additionally, the quality of the seams can be verified by a post-process ultrasonic inspection system immediately.

5.2. Process development for laser GMAW hybrid welding with high power laser beam

In addition to conventional connections like butt and T-joints, also examinations on new connection geometries, for example the combed T-joint, were implemented. The fundamental feasibility of the welding tasks is proved in the whole test series. Compared to other investigations on this technology with laser power less than 12 kW, it was necessary to adjust the process parameters with most significant changes in the focus position and the GMAW parameters.
Under use of the available laser beam power it is now possible to join typical material thicknesses of the shipbuilding industry in full connection from one side. Figure 5 shows a micrograph of a T-joint welded with a laser beam power of 18 kW.

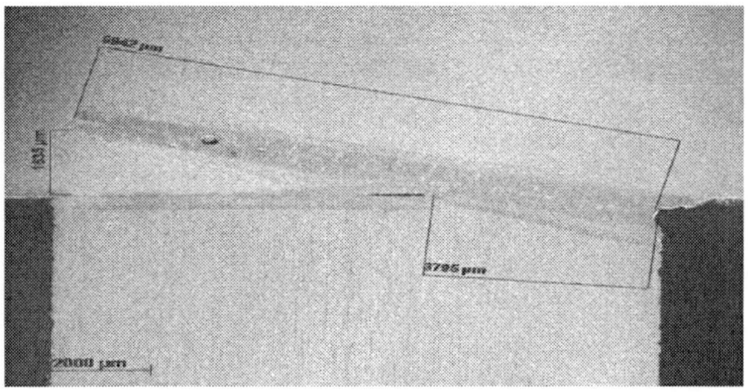

figure 5: micrograph of a T-joint, material thickness t = 10 mm [source: iLAS / Carl Cloos Schweißtechnik GmbH]

It is clearly seen that the welding depth is not the limiting factor, but the angular position of the optical processing system to the workpiece. In this case, the optical system was positioned at an angle of 15° to the workpiece so that no penetration occurred. It follows that an angle of incidence of less than 10° is required for laser only as well as for the laser GMAW hybrid process. This requirement found its way into the construction specifically designed for laser beam power of 30 kW and the requirements of the laser GMAW hybrid head in figure 6.
With this developed construction by the iLAS and Carl Cloos Schweißtechnik GmbH, it is possible to couple the laser beam flat to the workpiece and thus to achieve a high penetration depth. A to the laser beam coaxial rotatable additional axis for the GMAW blowpipe allows to position the blowpipe to the optimal angular position for the process. Two aspects were examined in the current test series. First, the accessibility was tested with the new head both by simulation and also in practice. Especially when welding the walls on the panel resulting inaccessible welding positions requiring multiple design revisions of the machining head to make this compact. At the second aspect the bridging and the seam tracking was in the foreground. The first experiments show the potential to bridge a gap of up to one millimeter, a gab, that is realizable in the future with the new design solutions in shipbuilding. For seam tracking it is developing a technical solution within the project based on the researches conducted accessibility testing, which allows a coaxial seam tracking.

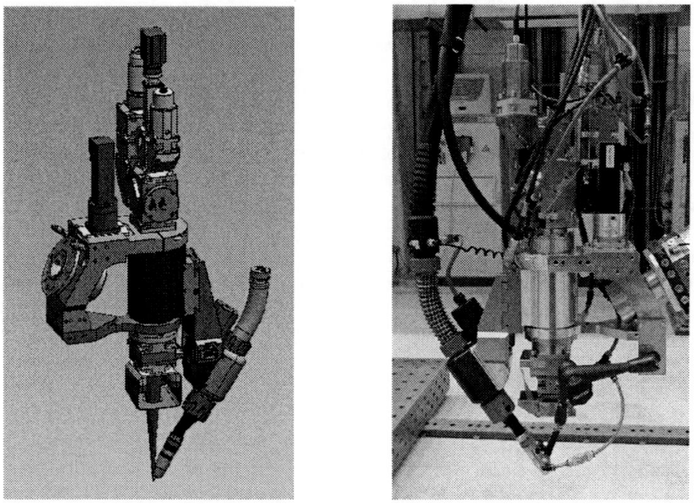

figure 6: laser GMAW hybrid head with additional rotary axis for the GMAW blowpipe [source: iLAS / Carl Cloos Schweißtechnik GmbH]

6. Construction of the robot gantry system

To proof the research results on life size ship structures a manufacturing demonstrator with three gantries was set up at the LZN Laser Zentrum Nord GmbH in Hamburg, Germany. By this way it is possible to perform experiments on structures with typical shipbuilding dimensions. The QuInLas manufacturing demonstrator is shown in figure 7 and has a length of 30 m and is capable to handle components with dimensions of 10 m x 3 m x 5 m.

figure 7: view of the manufacturing demonstrator with gantry system 1: laser remote welding system (front), gantry system 2: laser GMAW hybrid welding system (middle) and gantry system 3: ultrasonic inspection system (rear) [source: iLAS]

The workpieces are put together under the gantry with a 40 ton crane. The positioning is carried out on the component structure by joining the parts with the described concepts or with adapted clamps. After assembling the workpieces an integrated camera system in the laser remote welding demonstrator, mounted on the gantry system 1, measures the position of the components and seams. The data is used to correct the machining path for the welding process. In a first step the construction will be stitched by the laser remote process, so that the structure has sufficient stability for the subsequent welding. With gantry system 2, the component will be full welded with the laser GMAW hybrid technology. An online process diagnostics is accompanying the process and integrated into the machining head. Possible detected defects are further investigated via an ultrasonic inspection system mounted on gantry 3. The order of all preparation, manufacturing and quality insurance process steps resemble the sequences that have been proven fully practicable in commercial ship manufacturing.

7. Integrated quality assurance

7.1. Online quality diagnosis for the laser GMAW hybrid welding

The laser-based joining processes laser remote and laser GMAW hybrid welding need online-diagnostics to meet future requirements for three-dimensional welded constructions in shipbuilding. Because of its low thermal process induced distortions, advantages of the laser-based joining procedures become more important. They are used for the production of thin wall three-dimensional structures with thicknesses of t = 4 mm to t = 10 mm. The detection of the joint and process quality takes place under realistic conditions using the manufacturing demonstrator for flexible laser remote and laser GMAW hybrid welding and an integrated quality diagnostic and assessment.
During the laser GMAW hybrid welding intense radiation across the entire optical spectrum (UV, VIS, NIR) is emitted from the process zone. The intensity of the radiation varies with the pulse frequency of the GMAW arc. For optical process monitoring camera and spatial integration sensor systems are used. The laser GMAW hybrid welding process will be observed and monitored with the Coaxial Process Control System (CPC system) of the ILT with a high-speed CMOS camera by a narrow-band, optical bandpass filter. This observation will be done coaxially through the focusing lens. A super pulse diode laser, which is disposed laterally adjacent to the focusing lens and is synchronized with the exposure of the camera, delivers 'flash lighting' to the process zone. The moment of data acquisition is adjusted to the current pulses of the welding current source. The aim is the automatic joint tracking and the online monitoring of the stability of the laser GMAW hybrid process for joining three-dimensional shipbuilding structures. After each current pulse of the GMAW process a pulsed illumination samples images periodically. Out of the recorded images the relative position of the capillary (keyhole) with respect to the joining gap can be detected. Figure 8 shows a single frame of a CPC recording sequence which was recorded at the laser GMAW hybrid welding process synchronously with the current pulses of the welding power source.

figure 8: CPC recording sequence during laser GMAW hybrid welding which was synchronized with the current pulses of the welding power source. [source: ILT]

In the image on the right side of the laser induced keyhole between the welding wire and the melt a flying drop can be identified, which was melted of from the filler wire 2 ms before. In a stable process, it is repeated periodically with every current pulse of the welding power source. Due to the strobe effect the drop always seems to float at the same position. To observe the position of the laser beam relatively to the joint and automatically compensate deviations for example caused by a heat distortion of the components, a real-time image processing joint tracking control system is developed and integrated into the manufacturing demonstrator. This monitoring set up is expected to deliver valid statements regarding the quality and stability of laser-based joining process which can be used for application in the shipbuilding, steel construction, vehicle construction and in the manufacturing of pipes.

7.2. Non-contact ultrasonic diagnostic of typical shipbuilding welded connections

The quality of a seam is characterized by its shape and geometry, but also by macroscopic defects, such as pores, cracks or lack of fusion. Especially in highly strained areas, such as in various modules of a ship, the accuracy of the seams is of great importance. Within the project QuInLas a special ultrasonic-based inspection system helps to find these kinds of defects.
The adapted test equipment is applicable for error detection and characterization of possible defects in the seam. Because of the special design of the ultrasonic measure head, this allows an inspection of the seam, based on diffraction and reflection at the same time. A differentiation of in pore-like, spatially very limited and crack-like, spatially extended seam errors is possible. A close scanned inspection of the seam can be performed, when the test head is guided parallel to the seam during the test. Furthermore a depiction of current ultrasonic signals is saved as an A-picture and the checking system has modules that display B-pictures and local amplitude curves as well. To facilitate automated ultrasonic testing after the welding process, electromagnetic ultrasonic converters (EMUS) will be used. These converters are characterized by the fact that the ultrasonic signals can be stimulated and tapped on electromagnetic interactions coupling free directly in the metallic measure object, see figure 9.

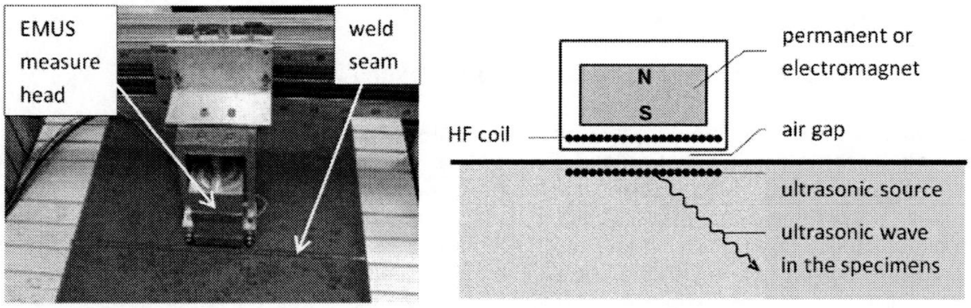

figure 9: ultrasound examination of an I-joint with EMUS measure head and laboratory manipulator (left) and a schematic construction of an EMUS measure head (right) [source: IZFP]

The obtained results show that, unilaterally unwelded and inhomogeneous areas of the seams can be reliably detected. Test errors, like blind holes and slots demonstrate the sensitivity to the relevant weld defects in T- and I-joints, see figure 10.

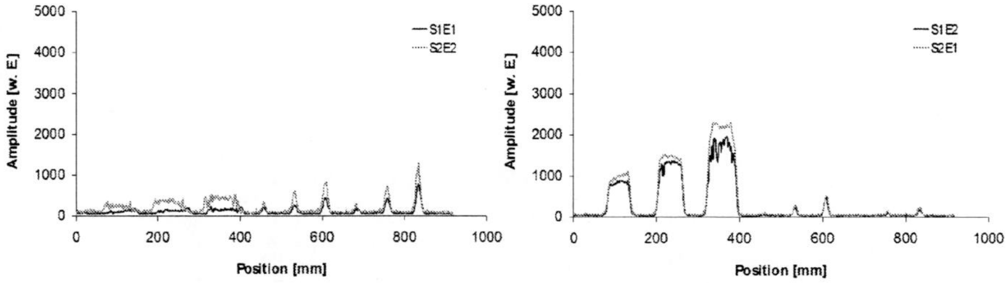

figure 10: local amplitude curves of the diffraction arrangement (left) and the reflection arrangement (right); order of the test failures: slots: 50 mm long (3x), blind holes ø 2 mm and ø 1 mm, depth is 0.5 mm, 1 mm and 2 mm [source: IZFP]

Furthermore it is possible to detect seam errors which cannot be found by classical x-ray radiographic examination. A subsequent destructive investigation of the corresponding cross-section of the seam confirms the results of the ultrasonic inspection.

8. Summary and outlook

The investigated technologies in the research project QuInLas meets increasing demands of the shipbuilding industry concerning the achievable production quality and time. For this purpose the laser remote welding process, which is well-known from the automobile industry, is applied to the requirements of the shipbuilding industry and adapted to the material thickness with matched laser beam power. In addition to the previous results, the use of this technology in combination with developed innovative ship designs let expect a significant leap in productivity.

The parallel online process monitoring of laser GMAW hybrid welding furthermore represents a new development that may facilitate and accelerate the quality assurance in shipbuilding. A comparison of the in-process quality diagnostics with the subsequent ultrasonic inspection allows a characterization of the seam error regarding the defect size. The influence of these errors on the seam strength is known from the performed fatigue tests. Thus, the inference is possible to determine whether the error lead to a detraction of the component strength and a rework is required. In addition, a comparison of the analysis results causes a reduction of misdiagnosis that may have an ejecting of OK components.

The developed solutions by the institutes and industrial partners are brought together in the described demonstrator and the technologies will be validated on real shipbuilding typical components such as section assemblies.

9. Acknowledgement

We thank the BMWi and the PTJ for supporting of the research project QuInLas!

Modern MIG/MAG – technologies using alternating current and their applications

M. Wege, Herborn/D, and K.-P. Schmidt, Haiger/D

Introduction

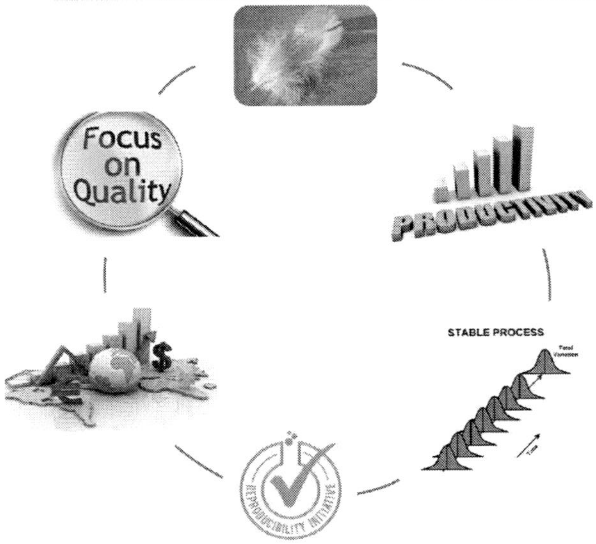

demands on welding processes

-lightweight construction
-Increase in production
-process stability
-reproductability
-economy
-quality

Alternating current-driven technology

Impulse driven process (DC)

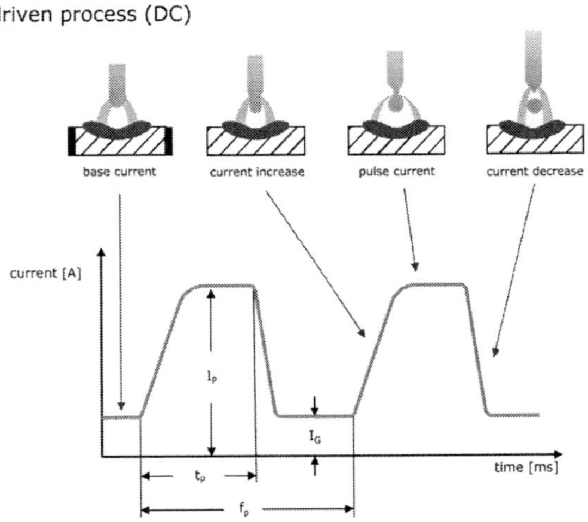

Alternating current-driven technology

Current-driven process (AC)

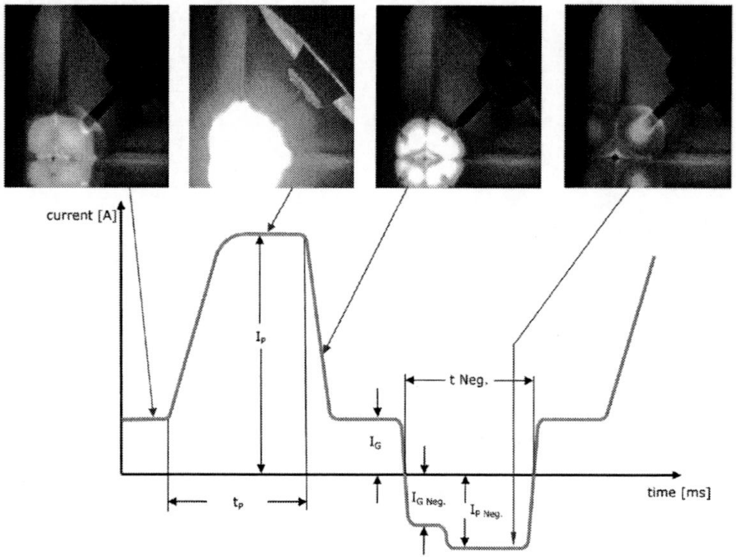

Alternating current-driven technology

Alternating current-driven technology

operating concept

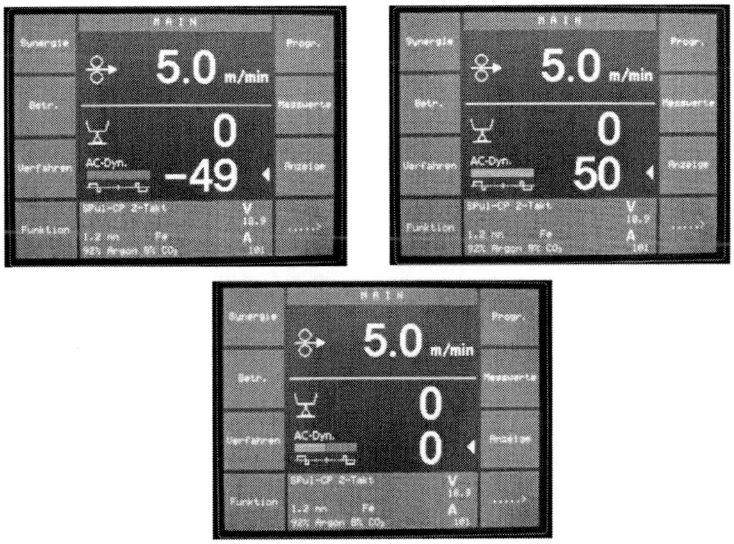

Applications

high strength fine grain construction steel

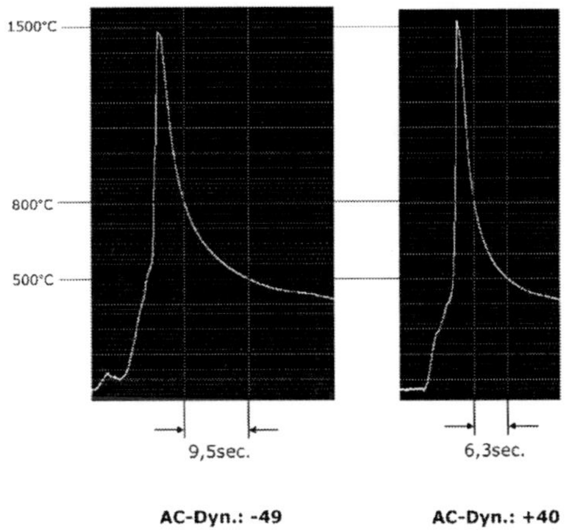

Welding parameters:

v_D:	8,5m/min.
v_S:	53 cm/min.
SZW:	Ø1,2mm; FK 1000
GW:	S1100QL
T_V:	110°C
Gas:	82%Ar / 18%CO_2

Weld buildup

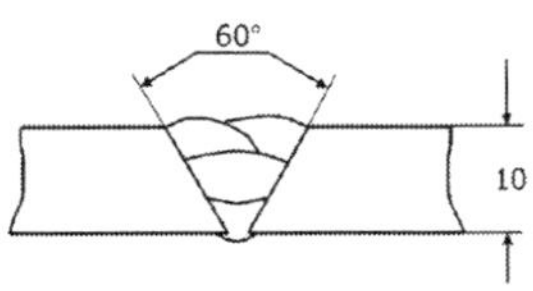

Applications

Aluminium

Fuel tanks for commercial vehicle industry

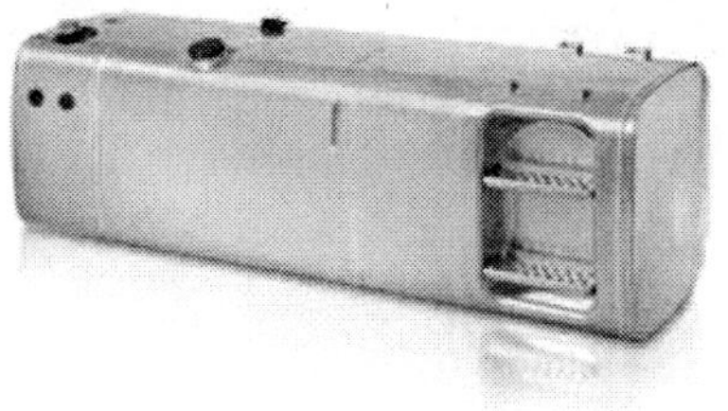

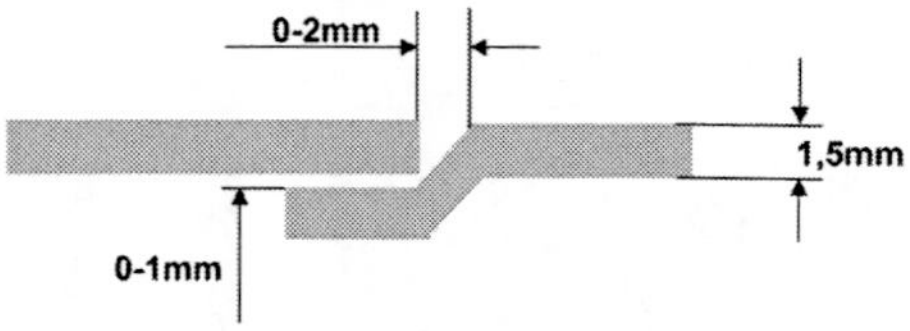

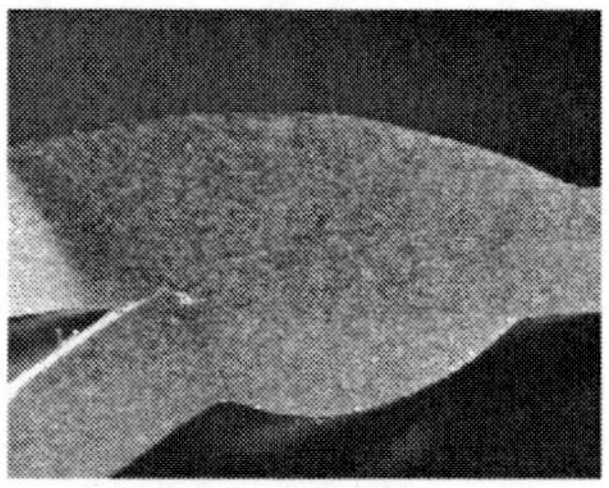

- **to bridge the vary gaps**
- **full root penetration**
- **tightness**
- **increase the welding speed**

Applications

Aluminium

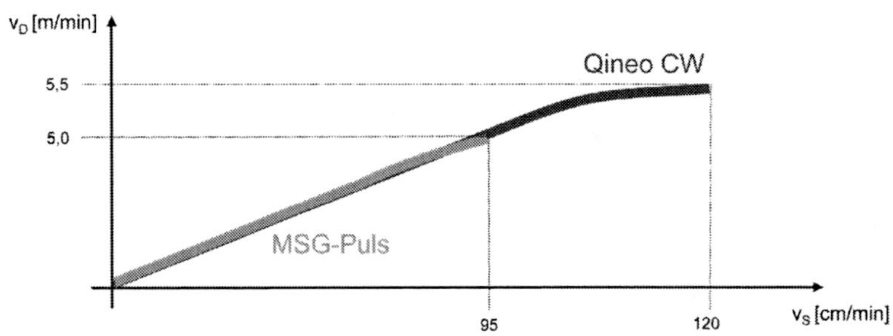

Welding parameters:

wire:	Ø1,6mm; AlMg4,5Mn
Base mat:	AlMg3
gas:	100%Ar

Applications

Aluminium

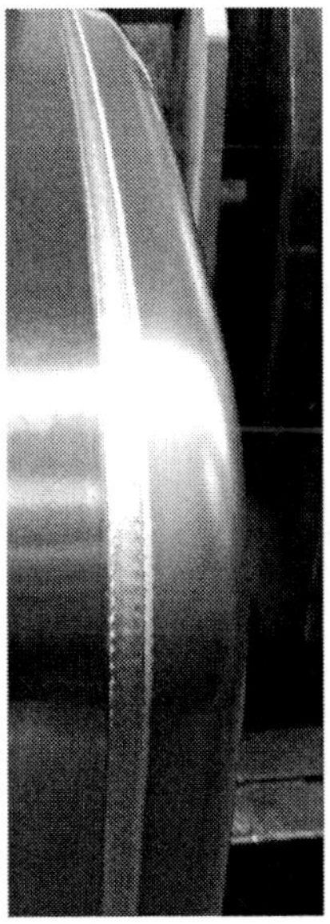

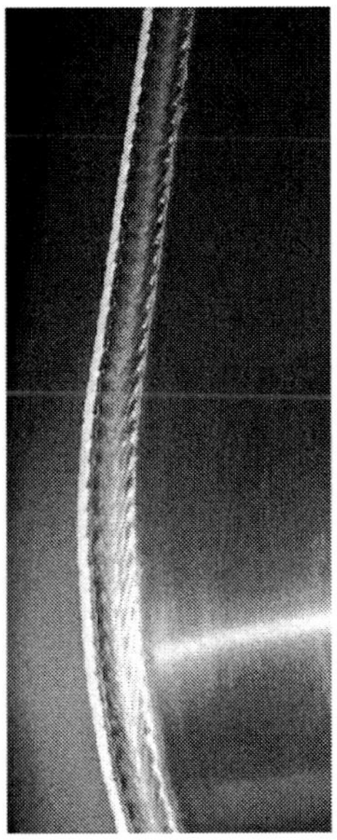

Applications

Stainless steel

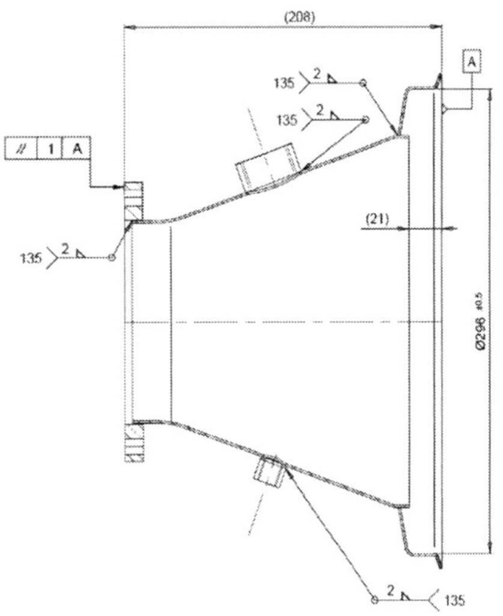

Applications

Stainless steel

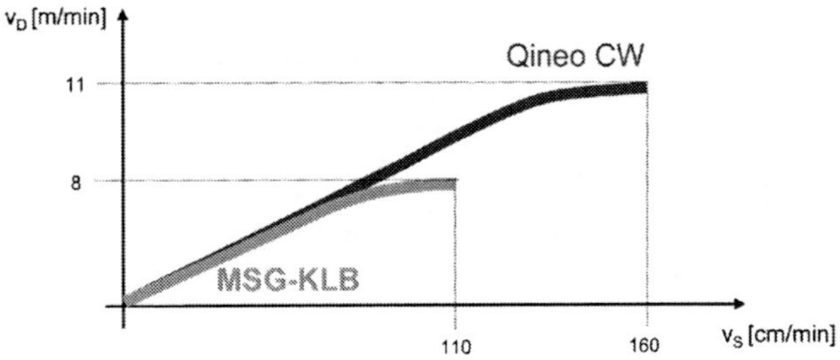

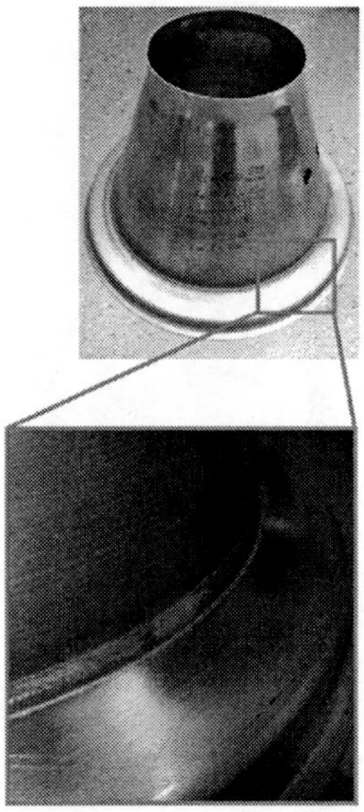

Welding parameters:

wire:	Ø1,0mm; 1.4370
Base mat:	1.4301
gas:	97,5%Ar / 2,5%CO_2

Summary

Automated weld seam inspection and inline seam optimisation

M. Maurer, Wiesbaden/D

Automated welding in industrial production is state of the art in countless industries. Yet in nearly all cases, the quality of the weld seams is still assessed manually by trained employees. This requires a large human resources contingent, while at the same time harboring the risk that poor-quality parts could remain undetected due to unavoidable human error.

To optimize production processes and enable consistent quality of weld seams, automated seam inspection systems have been used in various industries since the mid-1990s. In the automotive industry, for example, 3D machine vision systems enable automated weld seam inspections on axle supports and the chassis, on steel wheels, seats and exhaust systems; as well as inspection of heavy-duty shelves the supplier sector.

This presentation describes how automated 3D seam inspection systems simplify quality controls, ensure a consistently high quality of welding seams, and help to optimize welding processes and cost efficiency. And it will show how weld seam inspection system VIROwsi replaces expensive manual reworking with cost-effective, automated reworking of seams.

In 2007, Daimler AG became the first vehicle manufacturer to utilize a system for the automated inspection and optimization of welded seams. The overall concept developed by Daimler was implemented for the first time for rear axle carrier assembly on the C-Class. The goal of Daimler AG was to replace the existing manual welded seam inspection and reconditioning, with an automated inspection and optimization system working in a line cycle.

Efficient production of large components using a novel 3D FSW head

Th. Weinberger, G. Figner, Graz/AT, P. Freigassner, S. Hampel and F. Hampel, Obdach/AT

Friction Stir Welding (FSW)

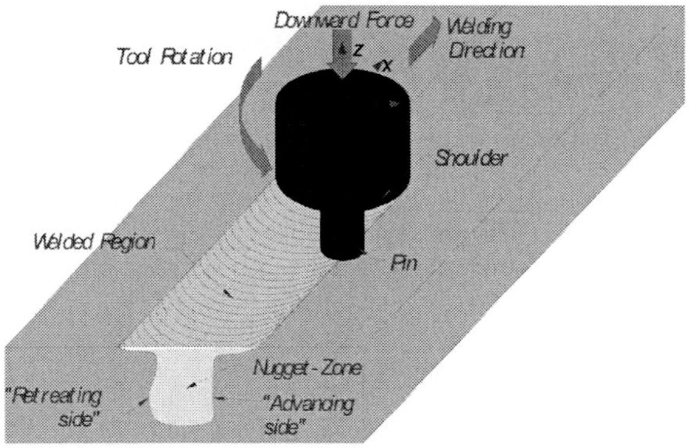

Large scale production

Example: 16m long part, many different processing steps necessary

Machine technology

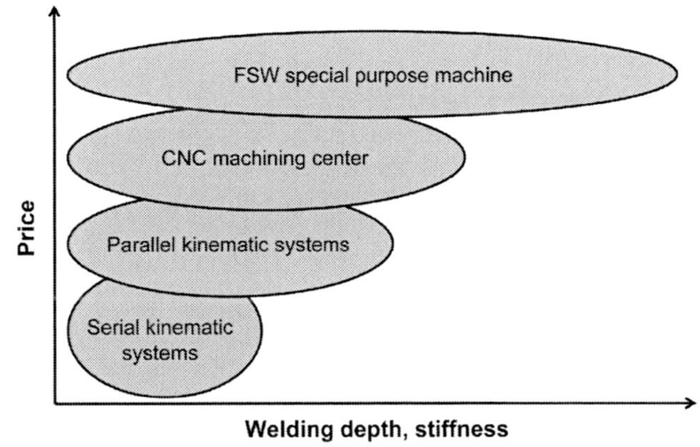

Machine technology

Machine technology

Friction stir welding vs. milling

Use of milling machines for FSW?

- Spindle → spindle speed, torque, bearings
- Drive, control system → welding speed, axial force / mode, programming
- Guides → loads, clearance

The use of milling machines for FSW is limited

Goals

Development and construction of a 3-axis machining head

•*FSW welding of components*

- For all common metals such as (high strength) aluminum alloys, copper, magnesium, mixed compounds (Al, Al-St, Al-Cu, Mg-St)
- 3D joint geometries

•*Complete machining operations* (milling, drilling, ...) of components in the same clamping device

•*WeldCheck* – Control software for FSW

Layout welding head

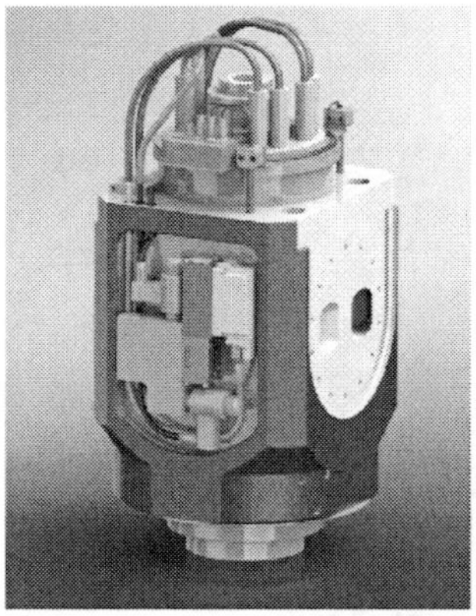

StirPower 300

✓ Hybrid head for
- Friction Stir Welding (FSW)
- High speed cutting (HSC)

✓ Position and force control mode

✓ Max. spindle speed 15.000 rpm

✓ Max. axial force 25kN

✓ Joint tracking system

✓ Automatic tool changer

StirPower 300

StirPower 300

FSW of aluminium extrusions

FSW with seam tracking device and rollers

StirPower 300

Application butt joint

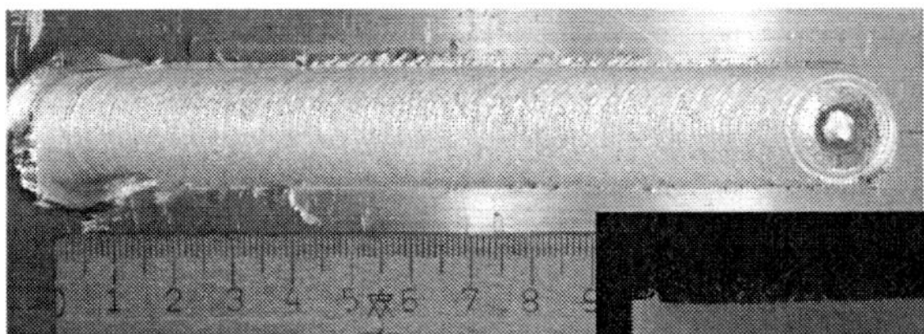

FSW of 15mm thick aluminium alloy EN AW6082T6

HAGEMATIC 305 FSW

Integration of Stirpower 300 welding head:

•Gantry machine with 6 NC-axis (4 linear axis, 2 rotary axis) for approx. 14 m long aluminium profiles, 3 processing stations, 5-side processing

- The HAGEMATIC 305 FSW is a 3D Friction Stir Welding machine (5 axis interpolation) as well as a 3D milling centre

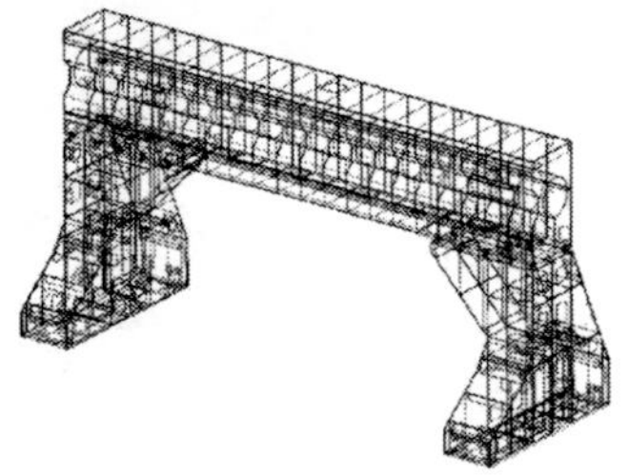

Hagematic 305 FSW

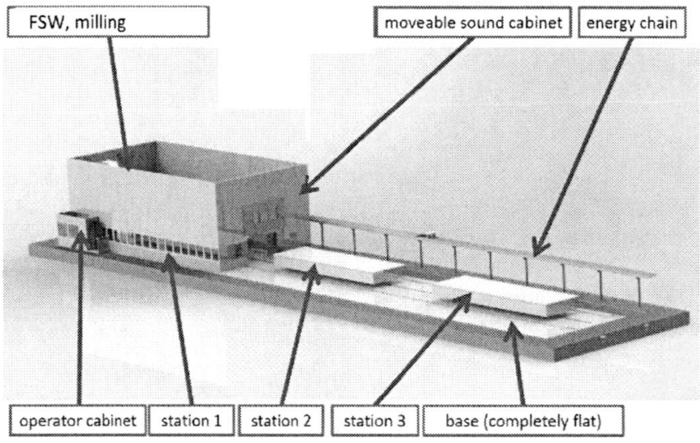

Advanced application

Use of multipart tools (bobbin tools, adjustable pin tools, ...) is possible

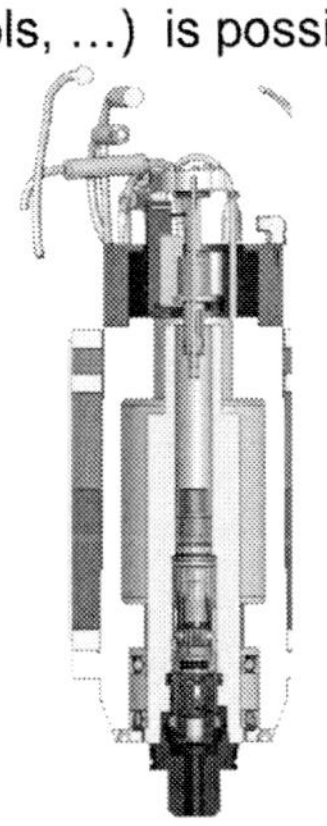

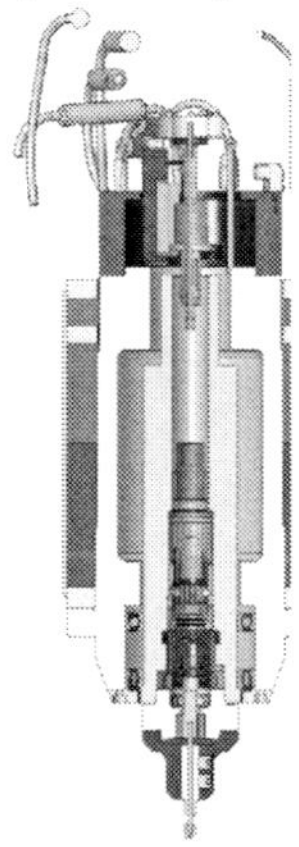

Adjustable Pin Tool

Friction Stir Welding of 15mm thick aluminium alloy using an ajustable pin tool

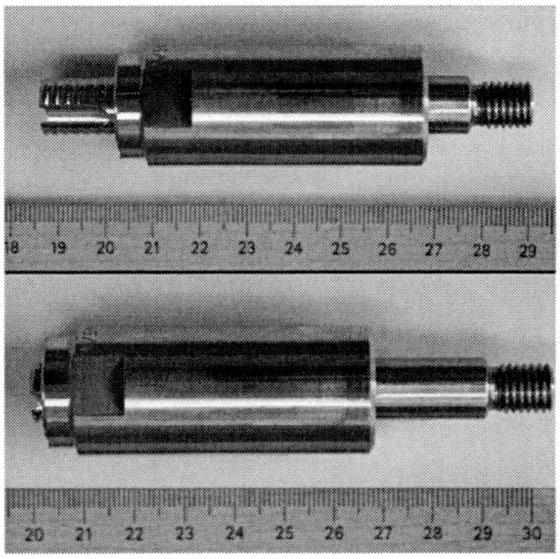

Adjustable Pin Tool

Advantages
•Variation of welding depth
•Elimination of end hole

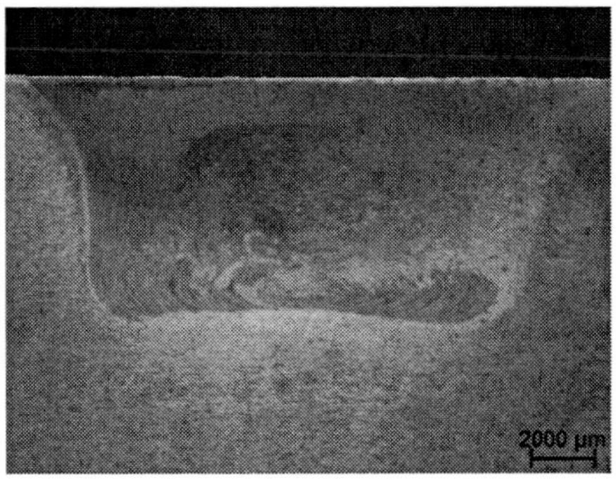

Variations machine technology

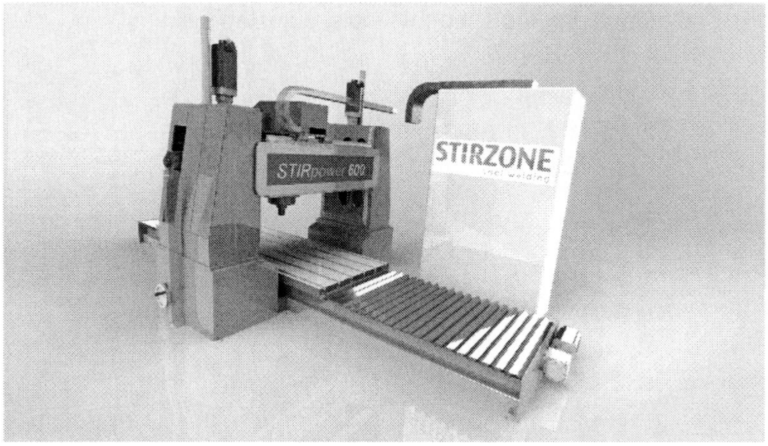

Variations machine technology

Key accomplishments

- Complex components can be processed using different machining processes
- The hybrid system can be equipped with various tools such as milling, drilling, cutting or sawing tools and changes them automatically
- Dual design of the system → lower non-productive time
- Reducing investment risk and quicker return
- The user is able to handle a wide range of jobs

Partner

HAGE Sondermaschinenbau GmbH & CoKG

Institut für Werkstoffkunde und Schweißtechnik
Technische Universität Graz

Science Park Graz GmbH

Österreichische Forschungsförderungsgesellschaft

Spot welding and gluing of structural parts in automotive industry

Th. Buschhaus, Obernburg/D, C. Clark, Elgin/US, and S. Zhi, Kunshan/CN

Following are examples from volume models (VW Passat), premium vehicles (Daimler S-class) and models produced worldwide (Daimler C-class).

The systems realized worldwide comprise up to 8 processes in total, such as:

- MIG soldering
- Laser cutting and welding
- nut welding
- Sawing
- Gluing
- Hardening / gelling of the glue
- Spot welding
- stud welding
- Washing
- Interlinking
- Automatic depalletizing

Thus wide-ranging know how for various areas is globally in demand.

Breakdown

Typical structural components in car body manufacture such as B-pillar and longitudinal beam, as well as transversal beam and web part wheelhouse, are subject to steadily increasing requirements. Doing so, methods are combined even more, e.g. gluing and spot welding, in order to obtain large-surface connections with high strength and rigidity. High-strength, hot formed press parts, roll-formed profiles and standard components like Isofix bracket are also used in the safety relevant assembly units, of course.

Picture 1: Spot welding and gluing of car body components

System concepts

The realizes systems are based on standard cells. These are independent units that will be operated independent of each other later, and that can be reused universally.

Especially the geo welding station is a central module. Normally rotary tables are used here for right and left h.s. components with 2, 3, or 4 stations. If possible, only one geo-fixture is used per production stage, that means e.g. one 2-station rotary table for left and right h.s. components. For a 4-door vehicle or multi-stage processes 3 or 4 stations are also quite common.

Welding of the pre-manufactured assembly unit will then usually be done with parts handling (jigless welding). The handling robot can mark and palletize the components finally. Particularly long products will also be welded in simple fixtures or on rotary tables because of the required reach.

The pictures 2, 3, and 4 show corresponding examples.

Picture 2: Multi process system with spot welding robots and gantry robots combined for gluing and interlinking

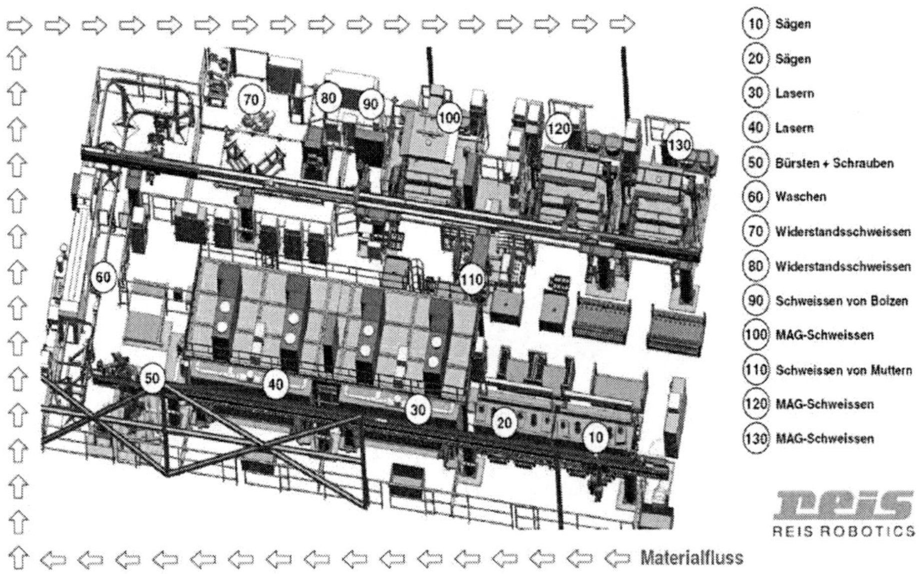

Picture 3: System with several standard cells and 8 processes

Picture 4: Spot and MAG welding robots in a system with 4 station rotary table and 5 processes

Gluing is normally combined with parts handling, too. Thus, the single parts can be inserted into the fixtures automatically without a worker's action. Often, parts are fed via a conveyor or a pallet so that the worker is uncoupled from the cycle. Furthermore, reasonably rivet nuts and other accessories can be mounted on the conveyor and sensed in the handover station.

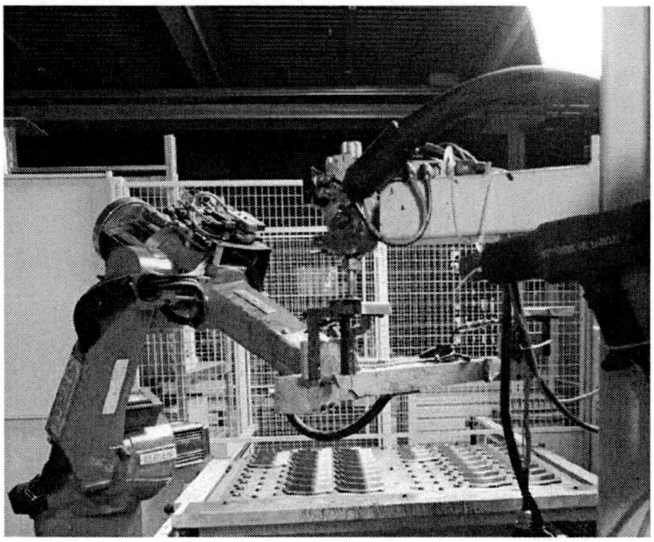

Picture 5: Parts infeed via conveyor or pallet and glue application

Finished parts are palletized e.g. separately as left and right components. Usually, load carriers are changed parallel to the main time. For this purpose, corresponding concepts with rotary tables or conveyor system are suitable. Loading of the load carriers is made by standard articulated-arm robots or, advantageous for large assembly units, by linear robots or hybrid kinematics.

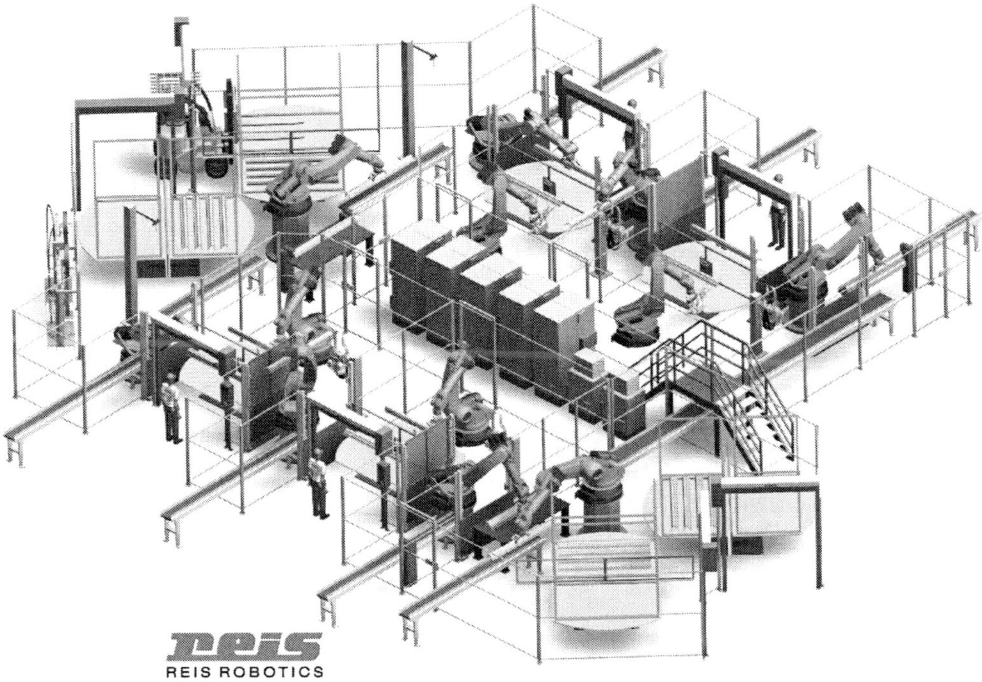

Picture 6: Loading of load carriers with articulated-arm robots and rotary table

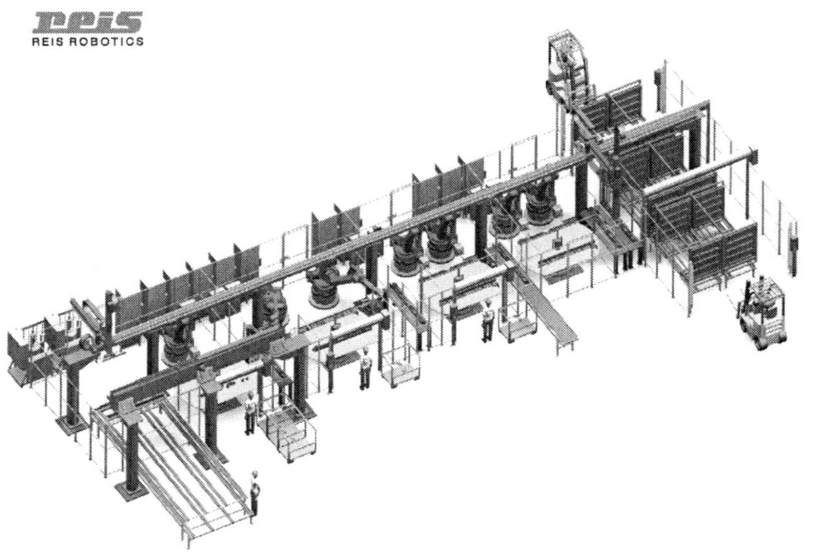

Picture 7: Overhead solutions for interlinking and loading of load carriers

Gluing technique

Both, single-package systems and two-package systems are used. Usually, this is chosen by the developer or the OEM/automotive manufacturer. Systems with automatic switch between two barrels are available for smooth, continuous operation. If needed, a heated follow-up plate can be provided in order to guarantee a constant temperature of the glue. The glue dosing systems made by different manufacturers work on different principles, e.g. with piston dosing unit and different pump designs. The glue nozzle itself can either be installed stationary or guided by the robot. The latter is preferably used for big assembly units.

When planning the work sequences it especially must be observed that the worker preferably does not come into contact with the glue. This serves for both, occupational health and safety and assurance of correct gluing.

Normally, the glue cures in the KTL coating at the automotive manufacturer. This means that the glued products can only be stored for a limited time and must be processed e.g. within 4 to 6 weeks. Thus the automotive component supplier and tier-one can only manufacture a limited amount in advance and must work strictly according to the FIFO principle. Relocations and modifications on the systems must be agreed exactly with the OEM. For worldwide supply of the assembly facilities, a worldwide production directly near the customer is necessary. Warming up the parts in a targeted manner is a possibility of improving the suitability for storage and transport. During this gelling process at typically 250°C for several minutes, the glue is pre-cured, so that it doesn't absorb moisture anymore. For this purpose, e.g. through-type furnaces or induction heating systems are available.

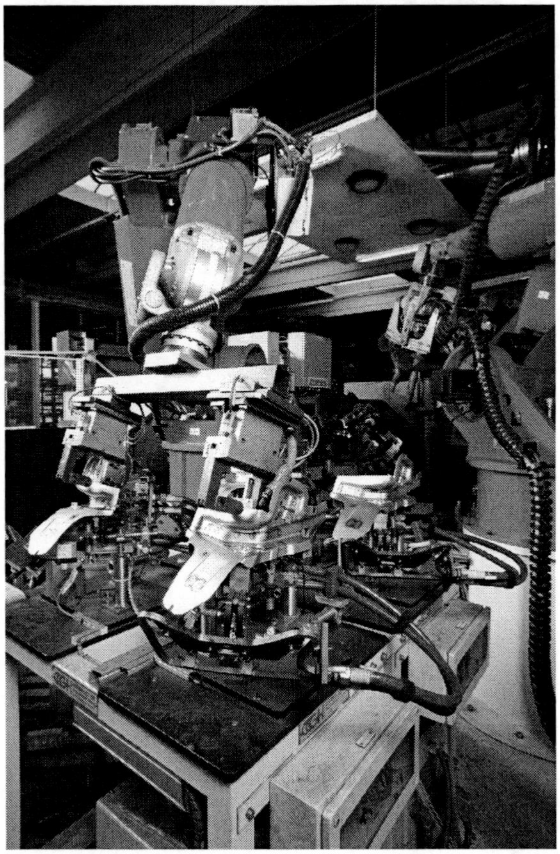

Picture 8: Gelling station

Welding-related challenges

Isofix-brackets made from steel, stainless steel and galvanized wire must be processed in a spatter-free way. For this purpose, meanwhile established standard solutions are available. For example, cold welding processes such as CMT, Cold ARC or Cold MIG are named.

The new challenges for both forming process and welding lie in the used high-tensile sheet metals which must be joined partly also as triple sheet welds. Considerably higher clamping forces are needed for this. The tongs are normally designed servo-pneumatically or, lately, driven by servo-motors. The servo-motor hereby can be programmed freely as 7th robot axis which partially leads to considerable time savings. However, as opposing tendency, longer hold times must be observed. The previously usual integrated pneumatic equalizing at the robot tongs is lately solved by the robot itself with corresponding software functions for reasons of cost and weight. The weight of servo-motor driven tongs without equalizing thus can be reduced considerably (30 - 50 %). This allows the use of smaller, more compact, faster and less expensive robots. Due to the higher rigidity of the products, usually stronger clamps which have to be considered mandatorily during accessibility examinations, are used.

Quality assurance and sensors

In arc welding as well as in resistance welding there is a clear trend to more and more quick process control. Doing so, it shall be ensured that the required welding parameters are actually reached. Nevertheless, regular control samples and destructive testing are still state-of-the-art.

Of course, during the gluing process an optical check before joining using gray scale or light section cameras is possible. A clean, constructive solution of the components and fixtures as well as the designing of the joining processes without the worker's influence is important. Doing so, it is ensured that the glue remains at the intended spots and e.g. is not sheared off. It is quite usual to apply the glue in such a manner that it is easily visible for the final quality check.

Services

Many customers work with
- local system integrators and
- create own concepts for the systems

But what happens, when globalization calls and one goes out into the wide world?

Good advice is in demand when space and budget are not sufficient for the wished system. Then, the world-renowned German engineering is needed. This includes e.g. good ideas for even better concepts, world-wide presence on site and, at the same time, one central contact for planning and coordination of the local and world-wide projects. The supplier of such integrated systems has to be able to analyze all processes, cycle times, worker times, capacities, time schedules etc. This is only possible with a powerful team of application specialists from the different named fields. For example, Daimler's C-class is produced in Germany, USA, China and South Africa. Since glued parts are not suitable for long transportation ways, the supplier of the B-pillar has to build works in overseas near the customer.

Reis Robotics has successfully installed movable systems of the same design which are structured in a modular way for these sites.

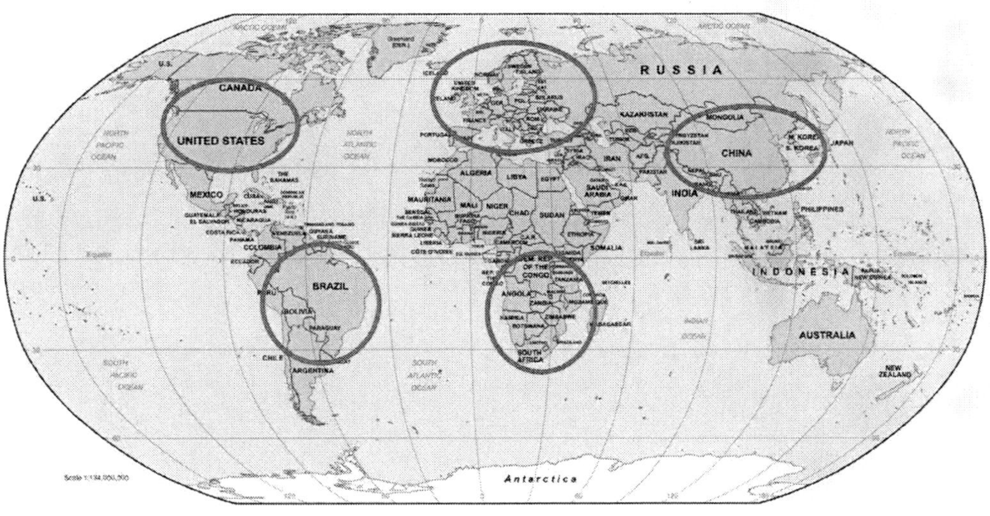

Picture 9: Worldwide projects require single engineering partner

Prospects

The combination of different production methods will, beyond the automotive industry, also arrive in other industries, such as manufacturing of control cabinets or in rail vehicle construction.

Picture 10: Spot welding still up to date

In the 1990s different prophets predicted, that pressure joining and laser will replace spot welding. Today we know, that the combination of gluing and spot welding e.g. enables advantageous large-surface connections and thus, these procedures are still up to date.

I dare to predict for the future, that also for structure components in the future more laser welding solutions will be applied, which today are still solved in another way due to reasons of accessibility.

The report during the congress will deepen these comments using a video film.

Picture credits: Rice Robotics

Aluminium joining: Friction spot welding and resistance welding – new possibilities for successful joining, experiences and applications

F. Luidhardt and A. Oelkers, Hamburg/D

Alumium is a demanding material with high potential in industry. Safe and reliably joining methods are a key for good usability in production. Both processes, friction spot welding and resistance welding offer safe, economic and reliable ways to join alumium and have made great progess in the near past.

Robotic friction stir welding of primary structural elements of a medium-size aircraft

J. F. dos Santos, A. Roos, Geesthacht/D, F. Fernandez, M. Cruz, A. Viliotti and M. Miyazaki,
São José dos Campos/BR

Aircraft manufacturers are continuously searching for ways to reduce structural weight and improve production technology. Friction Stir Welding (FSW) belongs to the group of green manufacturing processes and has shown to be highly suitable for the joining of high strength aluminium alloys. The potential for the use of FSW in airframes is extremely large and the benefits of replacing mechanical fastening or NC machining significant. It has been shown that for airframe alloys cost and weight savings of at least 10% can be anticipated compared to mechanical fastening.

One potential structural element is the joint of window frames to the fuselage. Conventionally this is a three dimensional riveted joint between a high strength rolled Al alloy (AA2024-T351) and a forged piece made of an AA7050 alloy. Within the scope of this project a welded solution has been proposed. Initially a process parameter window has been determined. Welds were produce to obtain samples for an extensive testing programme. This programme involved the determination of damage tolerance relevant mechanical properties such as fatigue and fatigue crack propagation. In the second phase of the project technology demonstrators were produced. This involved the manufacturing of clamping elements, robot programming and manufacturing the parts The friction stir joints revealed superior mechanical properties with values comparable to those of the base material. The welding speed proved to be twice as fast as the riveting feed rate. The complete structural element has been successfully welded by the Tricept 805 robot, consisting of: two skin/window frame joints, 6 overlap skin/stringer rows and a butt joint between the two sections. Based on these results, a 1:1 fuselage section has been produced and tested (barrel test) in large-scale testing facilities belonging to EMBRAER. The welded structured resisted 6,5 lives without detectable crack in the butt and lap joints.

Automation in welding – large robotic welding fabrication – high mix – low volume

D. Akey, C. Boyer, M. Davis and D. Rhoda, Fort Collins/US

Introduction

Wolf Robotics Chief Executive Officer Doug Rhoda will present high mix low volume capabilities in manufacturing. The following presentation and included manuscript delves into the model mix increases and production volume decreases that are high mix low volume manufacturing. High mix low volume is the production of a variety of products with small quantities in a plant, associated with make-to-order and engineer-to-order systems. Manufacturers have been challenged with issues of large component welding that will be further addressed; breakthroughs in technology have enabled automation to fabricate parts that were previously impractical. Offline programing, advanced robot carriers and multiple process capabilities deliver alternative solutions. These innovations have made robotic welding successful in high-mix low volume applications. Examples of new technology and sophisticated products along with case studies will be presented that show corresponding end user examples of successfully integrated manufacturing solutions.

Common Issues

Large component manufacturers are faced with a variety of issues including the understanding that "time is money". This includes the amount of hours required to gain that competitive advantage with the universal currency of time, of employees and work prosperity. Manufacturers are concerned with the training process of end users for new technology or processes that can be high level detail oriented or technical in nature. This cost of training employees or possibilities of hiring new talent that is familiar with the databases and adapt to the quick learning environment that comes with programming can be discouraging to many companies. Justification of part volumes also possess a discerning cost for manufacturers, the frequently too small volume is not sensible for the robots already in use, and therefore a purchase is not made. In comparison, the issues that commonly hinder manufacturer purchases with high-mix low volume systems require less supply chain discrepancies and open the door for new waves of improvement.

In the industry of manufacturing safety is a number one priority with ample preparation and repetitive licensing, companies must put safety at the top of the complications list. Machines, chemicals, environmental factors, tools or equipment and personal safety are all common issues needing to be addressed. Development of new technology lifts the safety burden with new processes and product quality improvements.

Quality and velocity difficulties struggled with by manufacturers become a hindering problem not only with daily functions but final production of units. The serving units of a system that adapts to velocity issues with product design engineering techniques can serve as a great improvement to prior systems or processes. Finding the correct intersection between quality and velocity can be a challenge for many HMLV manufacturers. High mix/low volume manufacturing is inherently volatile, meaning demand is often unpredictable and many items need to be made to the exact specifications of the customer. Designing a quality product takes time to do, but as mentioned before "time is money." This refers to the velocity aspect. The demand trend of today's economy is that products need to come quickly or it will cost the company money. If a manufacturer focuses on getting the product out fast, they may sacrifice the quality of the product and vice-versa.

Technology Advancements

Three-dimension simulation software allows programmers to simulate the necessary motions and processes of the robotics prior to start of production. Once the 3D simulation is completed, the programmer needs to code the motions into the "native language" of the particular robotics being used. Working in conjunction with the 3D simulation software, offline programing (OLP) allows you to program your robot from a computer rather than on the robot itself. Offline programming is the process of converting the sequence of events (created in the 3D simulation stage) and generating the robot program in the native language of the robot being used. Each robot manufacturer has a particular language that the robot reads and derives action from. Programming must match the native language of the robot determined or the robot can become vulnerable to risks. These risks include: the robotics not following the intended path, incorrect positioning, or possible collisions between the robots and other peripheral equipment and fixtures. When the 3D simulations of your part(s) and tooling/fixtures are imported, offline programming will then match that with the correct 3D model of the robot, creating a virtual replica of the system on the shop floor. After the virtual replica has been created, the programmer can begin sequencing robotic actions with a click of a mouse. To safeguard against the previously mentioned risks, simulation technology will allow you to run the welding program as a virtual simulation prior to actually using it. The simulation will play the programed sequence as written and even provide the completion time of the part. The software will show potential risks and errors with the sequence, and will allow for correction before implementation. After simulation and testing, the program is exported from the computer to the robot via Ethernet.

The integration of offline programming and simulation software drastically reduces the amount of time it takes to create and apply programs. Manual programming into the robotic cell itself can result in a large amount of idle time in which nothing is being created. As we learned earlier, velocity is key, and offline programming has been shown to take one-quarter of the time it used to take to program directly into the robots themselves. The biggest time savings comes from the ability of the programmer to begin writing a program for another part, while the robot continues to work on a current part. This ensures the highest amount of uptime resulting in a very high velocity in production.

Prior to offline programming, a separate program had to be written for every varying size of a single part, consequently resulting in time consumption and cost increases. A component of OLP, a macro, is a rule or pattern that specifies how a specific sequence should be altered to fit the varying sizes of the same part. A macro eliminates the need for multiple programs to be written for the same part varying in size and shape. The robot will take the macro and use adaptive welding practices called "sensing technology' to alter its path to fit the new part.

A single project or task can require a wide variety of processes. To accommodate this need, robots are manufactured with very high process flexibility. A single robot is capable of performing any combination of numerous processes, including: oxyfuel cutting, stud welding, plasma cutting, preheating, grinding, laser tracking, wire brushing, and scaling. These processes can be automatically interchanged, effectively reducing down time and manual labor. To offer the most flexibility throughout the entire system, the tool components of the system offer several adaptations and changes as well. For example torch necks can be interchanged to provide different types of torch access depending on the difficulty of reaching the required welding area/seam. The robot automatically switches between these torch neck variations according to the program sequence it is following. Process and tool exchanges take only a matter of seconds to complete, resulting in high velocity and reduced robot idle time.

Fixtures and Tooling

Fixture design strategy is important to the success of robotic welding in high-mix/low volume manufacturing. The goal is to present the robot with a consistent set of weld joint locations, while still providing maximum access for the robot to follow those joints. The more efficiently this is done the higher the quality of welds. There are several types of fixtures that need to be considered before beginning the weld.

A preliminary tack fixture is a fixture, mainly used for manual welding. Tack fixtures are a simple configuration designed to hold several parts in place with a series of small minor welds. Welds utilizing a tack fixture are not complete and only serve the purpose to hold parts in place for further welding. The holding fixtures are designed in conjunction with the tack fixture to add reinforcement to the lightly welded parts. Due to decreased clamp interference, along with increased mobility of a part, fixturing can offer access of welding both manually and robotic welding systems. Increased welding access can severely cut down weld time by increasing the efficiency of the welder. A single fixture can accommodate a variety of parts and sizes providing a universal "one-size-fits-all" fixturing solution. This eliminates the need to continually revamp fixturing to accommodate another piece, and greatly reduces the downtime associated with redoing fixtures. Fixturing allows welding projects to be much more efficient by adding simplicity. Fixturing can make the part easier to build, troubleshoot, and maintain because every angle of the piece can be reached and inspected easily.

Another simplistic tooling option is a system that works in union to securely hold the weldment in place and allow for maximum manipulation and maneuverability without complicated and expensive fixturing. This type of feature securely grabs and holds the weldment to allow optimum rotation for processing. Its quick change capability permits easy interchanging between other preloaded and staged weldments.

For more complicated and sophisticated components, tooling can be configured from simple manual toggle clamping to a pneumatic or hydraulic operated system with several degrees of safety related measures that prevent accidental release of components.

The final step in this process is the utilization of holding & locating individual components without the use of fixtures; a type of flexible fixturing that is mainly used for smaller parts. This variety of fixturing requires two or more robots working conjointly welding an object; one robot locates, positions and holds each individual part to be welded together while the second robot welds the parts together. This is specifically utilized in areas where numerous parts are in close proximity to each other restricting access of the welding torch while eliminating the cost of expensive and complicated fixturing.

Multiple Process Capabilities

Technology advancements in the capabilities that are associated with high-mix low volume systems and implementation improves time efficiency and productivity. Changing from manual to automated process requires the willingness of employees to learn and be trained alongside new systems to effectively gain maximum potential. Automation of tool exchange can improve a manufacturer's product quality and increase safety among employees. Torch capabilities such as preheating, plasma cutting, wire brushing and scaling contribute to the efficiency of the robotics, decreases time consumption for end production and expands the capabilities of the robot system. Cost effectiveness is a key success factor for manufacturing companies. For instance, plasma cutting and scaling are more efficient cost effective solutions than other standard cutting or cleaning systems. Abilities cohesive to these techniques have changed significantly in recent years, allowing for multiple materials to be cut, a longer consumables life and improved motion control capabilities.

Tool changers permit the use of numerous processes in the production cycle. Switching between tools is quick and easy and calibration of the tools is also automated to provide repetitive accuracy without operator involvement.

Processes not typically associated with robotics are now more feasible. Grinding operations were difficult to perform due to changes in wire wheel diameters, as they wore during operation, and maintaining the correct amount of pressure to accomplish the desired results. Now robots, together with new technology, have the capability to measure and maintain a prescribed amount of pressure, even as the wire brush or wheel wears down. Preheating of heavy metal parts can now be regulated and measured with sensors to ensure the correct temperatures are reached.

Torch neck configurations significantly impact all welding processes and are available in numerous configurations, sizes, angles and lengths to accommodate hard to access locations while maintaining the correct weld angle. A torch neck changer will allow the robot to interchange to the proper torch neck for the area being welded. An automated torch cleaner can keep slag removed from the nozzles, clip the welding wire to the correct length, clean the outside surface of the nozzle and spray a lubricant to prolong operation. Add a calibration unit to ensure the correct TCP of the welding torch and you prolong the automatic operating mode of the robot system.

Laser tracking, gas metal arc welding and gas tungsten arc welding use advanced processes that contribute to producing stronger higher quality welds. These processes grant the operator and robotics greater control over the weld and competing processes. A semi-automatic or automatic system deliberately reduces the time of production which then results in cost effectiveness. Gas metal arc welding is the most common industrial welding process for its versatility, speed and easy adapting process to robotic automation.

Advanced Carriers and Positioners

A single object can have a multitude of fabricating needs; the object must be manipulated and oriented to allow for all fabricating processes to be completed. This may require the use of carriers and positioning equipment capable of manipulating the robot or maneuvering weldments of 250-140,000 kgs in weight. These carriers and positioners must safely and quickly rotate and manipulate the robot or weldments to allow for increased access and processing.

Carriers include floor track systems and towers that can expand the working envelope of the robot. Floor track systems provide a means of transporting the robot along a horizontal plane to either move between stations and/or to extend the reach of the robot with large, long, heavy weldment parts. Robot towers permit the robot to be upright or inverted and utilize its full working envelope. The towers can also include the capability to raise and lower the robot, rotate the robot 360°, and even travel on the floor track system to further expand the robots reach.

Positioners are available in numerous configurations that allow for simple to complex manipulation of weldments using 1 or more axes of movement in coordinated motion with the robot. Incorporating the correct design is important and you must consider current and possible future needs along with the size and weight of the weldment and tooling together with calculations on moment of inertia. Proper counterbalance to compensate for the weight of the part or changes in center of gravity will ensure smooth operation and long life of the positioners.

The positioners can also incorporate a flexible fixturing system to allow for complete and total positioning of objects. This quick change fixturing system will permit simple loading/unloading of weldments onto related positioners using single or dual point location devices. See Section 4 for more details.

Customization

Vision technology advancements have greatly increased the reliability of robotic welding associated with high-mix low volume manufacturers, while still reducing the costs associated. Vision technology refers to the robot's ability to see every detail in the welding process. This ability greatly reduces the risk for defective products or manufacturing error. Vision technology would not be applicable without the utilization of laser technology. A low-power laser beam is projected on the surface of an object then reflected scattered light is imaged back onto a Charge Coupled Drive (CCD) or Complementary Metal-Oxide Semiconductor (CMOS) sensor. The laser is capable of detecting and identifying potential problems before, during, and after the welding process. The earlier a problem can be identified, the easier and less expensive it is. A majority of welding problems can be attributed to variability in the fixture or the parts being welded. This variability can cause the weld joint to vary in location resulting in both productivity and quality problems.

Laser tracking technology is a crucial part in the preproduction process. It can be used to track the joint, and measure the location and variance. This information can be used by toolmakers to adjust the fixture and/or move the part to close any gaps that may have been detected by the laser. This technology is so advanced that it can determine if the detail parts are being made correctly. The collection of this data can allow potential problems to be identified and resolved efficiently and effectively. Once parts and fixtures are optimized, the laser tracking technology will still be used to maintain proper torch position; the laser will continually scan and alter welding schedules while continuously adapting to compensate for increased variances. Weld quality will be increased by repeatedly altering the welding schedule to match the intended pre-programmed path. Post-weld laser inspection is used to ensure that the weld was done correctly, if found that a high level of variance in the weld, the laser will locate it and recommend solutions to increase that specific weld.

Laser tracking allows the robot to "see" the weld, while touch sensing is an adaptive welding practice that allows the robot to "feel" dimensions of the weld. Touch sensing is a low cost, software based joint locating system that gives the robot data to actively alter and adapt the weld to match the pre-programmed desired path. The robot uses a welding electrode, nozzle, or other sensing pointers to make the necessary electrical contact with the part. The robot then uses this data, along with algorithms, to make adjustments and automatically alter the weld path before the arc welding begins. This process creates an increased cycle time, due to performing search routines; but it can greatly reduce costs and time associated with fixing errors in the welding process.

Adaptive welding technologies such as these can prove to be very beneficial to the welding process. As mentioned before, the earlier a potential problem can be identified, the less time and cost it will take to resolve it. Adaptive welding technologies can greatly reduce costs by eliminating problems. Conclusively these technologies provide consistent quality products and high results in weld reliability.

Laser tracking is designed to withstand any welding environment and is not affected by ambient lighting or material surface conditions. Adaptive weld monitoring allows the robot to compensate during the welding operation for changes in weld location, gap width and seam movement due to the heat affected zone.

Man Machine Interfacing

Tooling advancements make substantial differences to manufacturing companies along with the technology connections between man and machine. Sensing technology can make machines smarter enabling the human connection but virtually eliminates human error. Seam finding, seam tracking, touch sensing and multi-pass welding can be useful applications that are process oriented. Welding interfaces such as cell control units allows a program that controls robotic movements displayed for the user to interact with the system. This specific connection of man and machine utilizes features and graphics that give operators complete and total control. Many manufacturing companies use similar systems that allow total control overview of the system and continual diagnostics of performance. These systems have a variety of features that can improve daily systems operations for example operator tracking, systems diagnostics and service functions. Man machine interfaces allow the operators to access a control panel for common service procedures without using the robot teach pendant, simplifying equipment operations. Versatility and progress are gained by utilizing man machine interfacing in high-mix low volume manufacturing which will then allow for total efficiency.

Case Study

Following is a case study performed by Jay Haynes Wolf Robotics regional manager presentation at the National Robotic Arc Welding Conference in the year of 2013.

"Robotics can enable high-mix/low volume manufacturing through the use of three factors: an enhanced ability to use multiple processes in a single cell, flexible fixturing, and improvements in software and offline programming. A single robotic cell can provide many processes including grinding, oxyfuel cutting, stud welding, laser tracking, wire brushing, scaling and plasma cutting. Alternate tools can be automatically changed by the robotics, improving cycle time. Flexible tooling is another method that allows manufacturers to take advantage of robotic automation. Wolf Robotics has developed several flexible tooling solutions, including SmartGrid, SmartDock and SmartClamp. The SmartGrid fixture plate is comprised of 1,080 threaded holes allowing for infinite tooling possibilities. Variable force SmartClamps are placed strategically on the SmartGrid, and are calibrated to allow the right amount of force needed to hold a specific piece in place on the fixture plate. Once the Smart clamps are placed and the piece is held securely the SmartGrid is then attached to the positioner using the SmartDock unit. These three components work simultaneously with the robot and positioner to manipulate the weldment allowing numerous processes to be completed on virtually any part. Parts of varying shapes, weight and sizes can now be secured by one fixture, thereby decreasing tooling costs and loading times."

Optimization of the variety of processes included with high-mix, low volume and appropriate tool usage results in high production levels and customer satisfaction.

Conclusion

As production volumes have decreased, production variety has increased. Manufacturers are assembling a greater mix of products than ever. One of the main reasons for this shift is that customers increasingly want products tailored to particular needs. High-mix, low volume allows companies to react to changes in the marketplace. The advancements of new technologies such as offline programming and macro-based programming contribute to a more efficient and effective ways to reach full potential. Specific features utilized to full potential can help manufacturers to become more advanced and capable of high productivity. Making the transition to high-mix, low volume manufacturing may appear as a complicated process but with proper preparation and acceptance of new technology advancements any manufacturer can utilize this new practices full potential.

Automation – world market, technical highlights and future evolution of the North American robotic welding and cutting industry

M. Oxlade, Auburn Hills/US

A presentation of the evolution of North American robotic thermal process over the last ten years through technology and methodology. A review of what has taken place and how it has lead to where we are today and the expectations of tomorrow.

The latest development of welding and automation in P.R. of China

W. Wang, Y. Shang and D. Sun, Changchun/CN

With the rapid developing of economy, China has gradually become a manufacturing giant in the world. Welding is one of the most important technologies of materials forming and processing in modern manufacturing industry.

Based on the overview of recent development in welding technology in China, the strategies for future advancement in the concerned fields of welding are proposed. On the one hand, in traditional manufacturing industries such as railway car body production, welding and automation should be further integrated to ensure the weld quality and consistency. On the other hand, considering the dominant role of iron and steel in the structural materials, welding metallurgy and high grade welding consumables for the new type metal materials should be promoted. As the largest electronic producer all over the world, the research and development of lead-free soldering materials and packing techniques are eagerly needed in China. The study and application of high energy density welding such as laser beam welding, plasma welding and electron beam welding should be encouraged to enhance welding quality or productivity. The numerical simulation of welding process such as thermal process, welding metallurgy, residual stress and deformation should be deeply investigated in future welding Industry. Finally, intelligent welding and control are becoming development trend in modern welding industry. A lot of research work needed to be done for China in this respect.

Use of automation & robotics in advanced welding systems in Turkey

M. Koçak, H. D. Gençkan and E. Erdem, Istanbul/TR

Abstract

Turkey is one of the most rapidly growing markets over last years regarding on robotic and automation systems used in fabrication. Turkish companies have started to employ robotic and automated systems in order to stay competitive in todays national and international market. As it is known that robotic and automated systems play an important role in achieving higher quality and productivity in a cost-effective way. Especially automotive, agricultural, railway industries and other fabrication industries related with welding operations increase their investments on robotic. Robotic welding and automation both provides continuous welding process with high production rates and is much safer than manual welding techniques in many applications. In this study, overview of the automation and robotic welding technologies used in Turkey will be summarized and some examples and their competitive advantages in their fields and applications of automated systems will be underlined.

INTRODUCTION

Robotic automation support is increasingly required by various industrial sectors such as automotive, agricultural, construction, transportation etc. in Turkey to meet the quality standards that are necessitated by global competition conditions and to be able to manufacture a large number of items. In Turkey design and manufacturing firms increase their efforts to provide new and innovative solutions in robotic and automation welding to enhance the benefits from welding robots.

Automated welding has become so popular because of the many benefits. These benefits include:

- **Increased Weld Quality-** Welding robots produce consistent welds, so every part you weld will be the same as the one before and after it.
- **Increased Ouput**- Welding robots work faster than humans and can work around the clock. This allows for an increased production rate.
- **Less Wasted Materials**- Since robotic welders are efficient and consistent, less materials are wasted due to poor quality or failure to meet standards.
- **Decreased Production Costs**- With welding robots manufacturers save on labor cost, since they do not require a salary. Welding robots also save on material costs since they waste less material.

In Figure 1, it demonstrates that estimated annual supply of industrial robots at year-end in total world by applications 2009 – 2012. It is seen that robotic welding applications is increasing day by day all over the world.

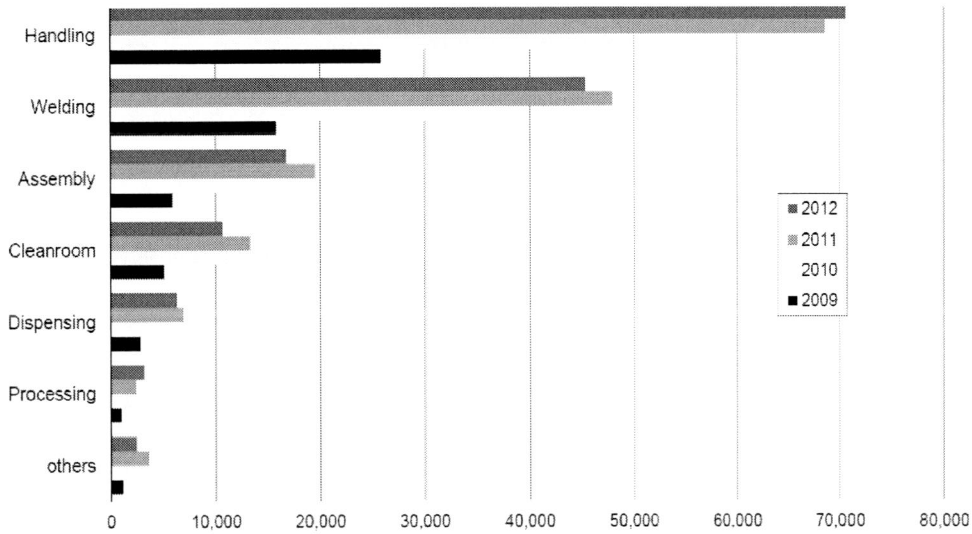

Figure 1: Estimated annual supply of industrial robots at year-end in total world by applications 2009 – 2012 [Source: IFR Statistical Department]

Some different applications of robotic and automation in Turkey are presented below.

RIM WELDING APPLICATION

Welding Fixture: This is a unique design for a combination of 50 different rim geometries and 3 different pinna geometries. Pistons on the inside of the fixture are forcing the pinnas to the inner surface of the rim and the pistons on the outside are fixing the rim geometry.

System: Used a slider for increasing the robots working space. Using a 2 stationed double axis positioner for welding the work piece and to eliminate the loading – unloading time.

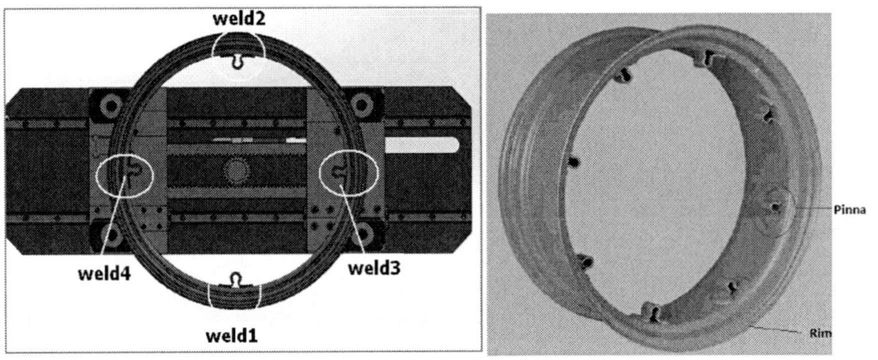

Figure 2: Welding regions and rim product

Rim materials are rim steel called S275 JRC and S235 JRC. These steels are cold forming and normalized steels.

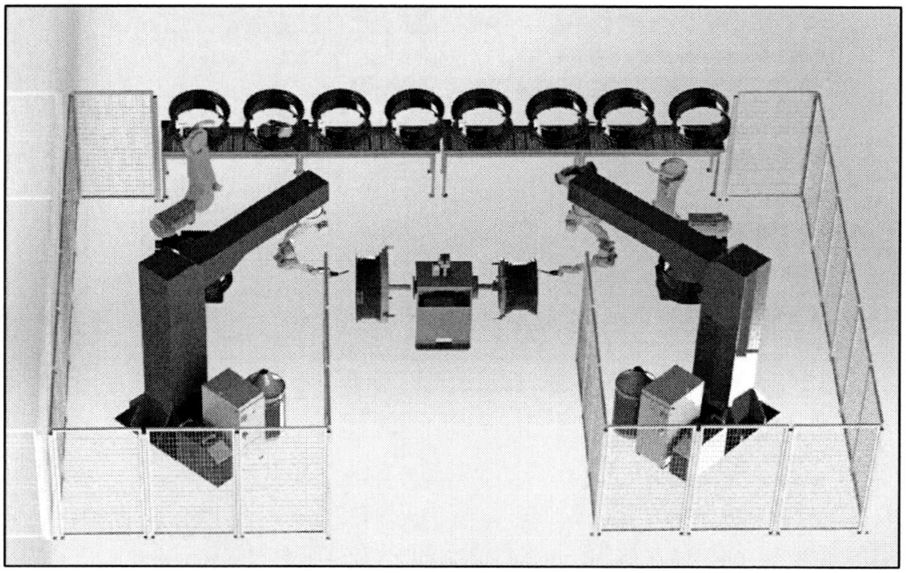

Figure 3: Design 1: Layout of the rim welding project

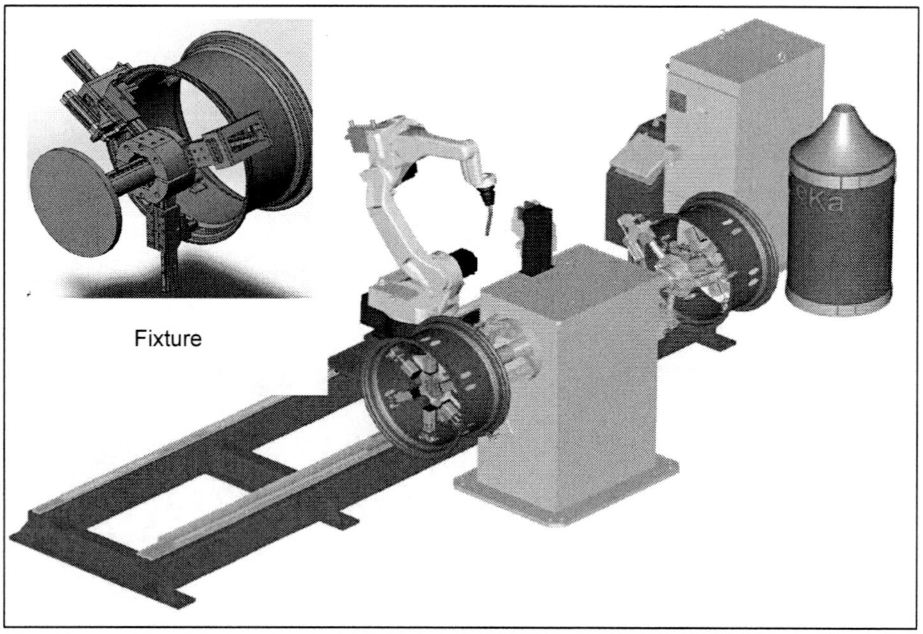

Figure 4: Design 2: Layout of the rim welding project

Figure 5: Robotic welding of the rims

BOGIE WELDING APPLICATION

The system basically consist of four main welding stations for vagon bogie, one station for assembly of the locomotive bogie and also two additional stations for blasting and Non-Destructive Testing.

In the first station, bogie side beams are welded. This process is realized with double pass manually, however when robots are employed tandem twin welding machines are used. In order to increase accessibility, robots move on a 9m slider and pieces are attached with pneumatic fixture.

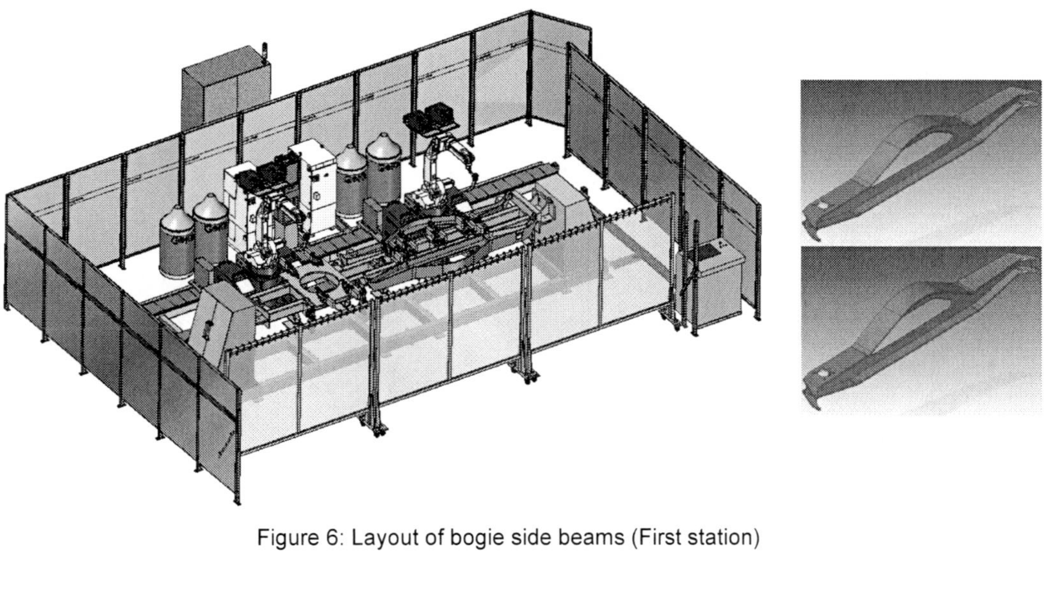

Figure 6: Layout of bogie side beams (First station)

In the second station, welding of mid traverse of the bogie is done with a similar robotic system used in the first station with the help of the slider and positioner. The semi-finished traverse and beams coming out of station 1 and station 2 are assembled in the 3rd station.

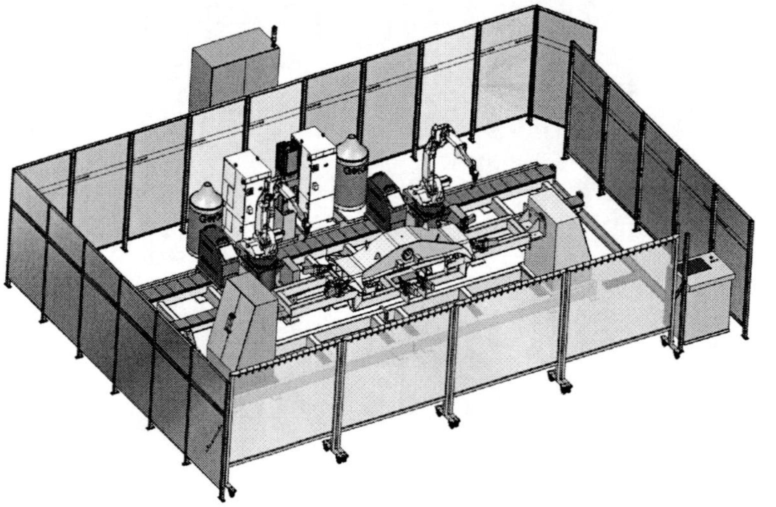

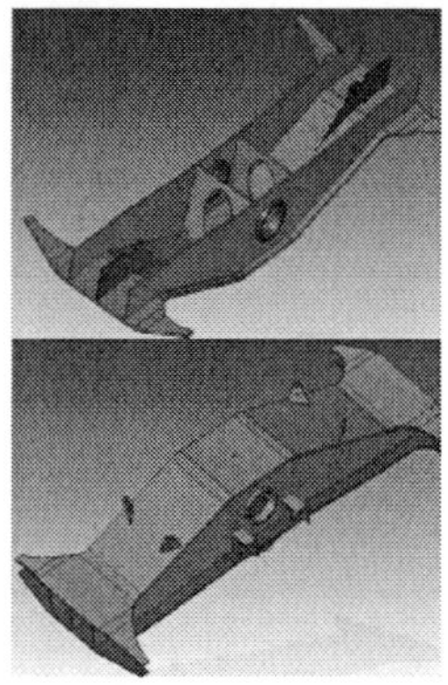

Figure 7: Layout of mid traverse (Second station)

In the third station, in order to increase accessibility, the robots which are hanged upside down are employed. Once the hardware and the robot trajectory programming is ready, welding parameters of the pieced are determined. All semifinished sub pieces' voltage, amper and welding speed parameters are monitored in the system. Thus, any faults can be detected by checking the date and parameters of the production. All the data regarding semifinished pieces stored in a compact way with the help of RFID technology.

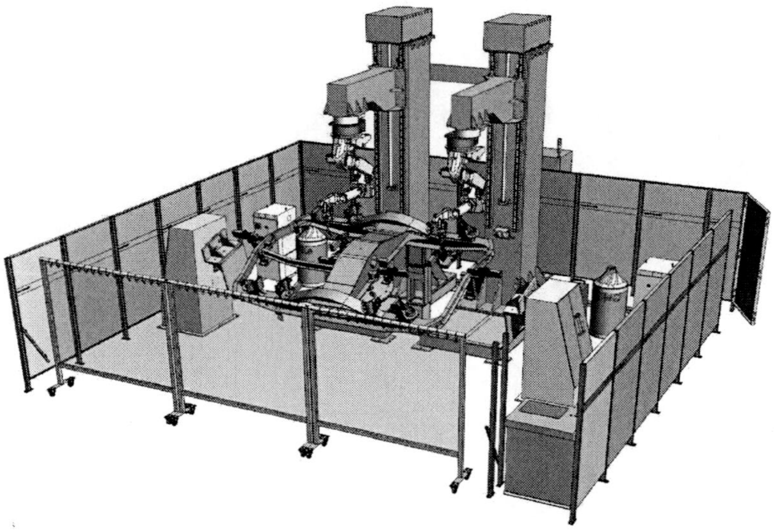

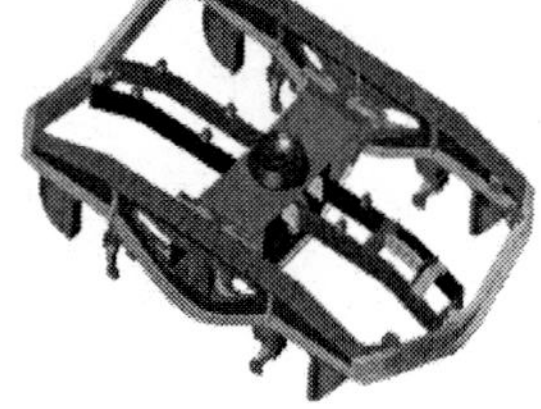

Complete bogie

Figure 8: Layout of plenary bogie (Third station)

In the fourth station, locomotive mid traverse, locomotive side beams and cast chassis are welded.

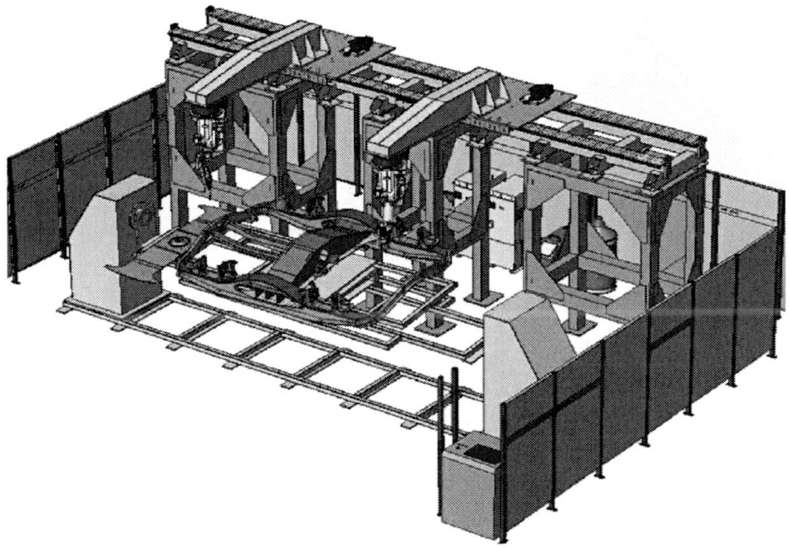

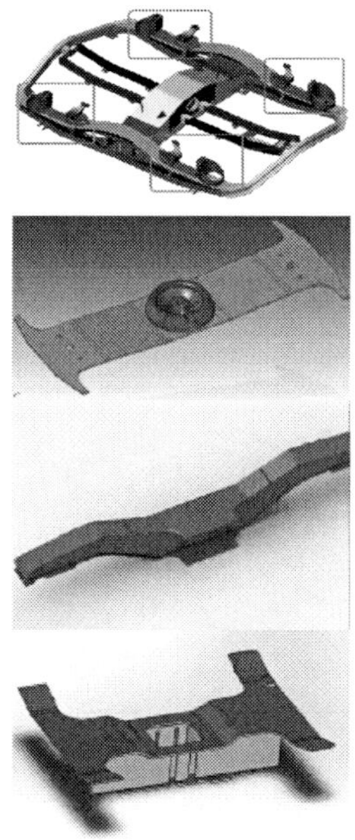

Figure 9: Layout of fourth station

At the last station, because of the waste of time of touch sensors for small size of welding areas, bogie is welded manually. Operators are able to recall programmes from touch panels and have control on the system. Fully welded vagon bogies are processed in blasting station in a way to catch surface quality norms described in ISO 8501-1 and all the welds in the system are inspected with phase array method.

AXLE TRUCK ROBOTIC WELDING SYSTEM

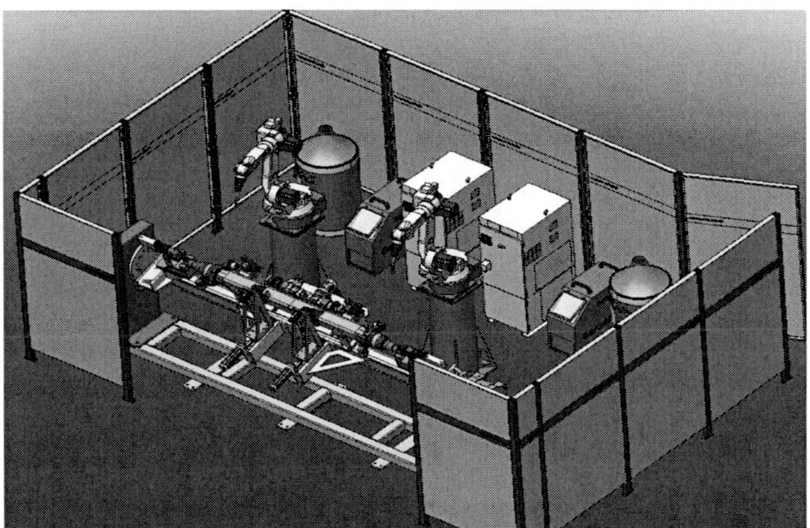

Figure 10: Layout of robotic truck axle welding

Robots acting as Twin, weld the axle of the truck. Welding fixture using the pneumatic clamps for fixing the axle and the 1-axis positioner helps to turn the work piece for a better welding trajectory.

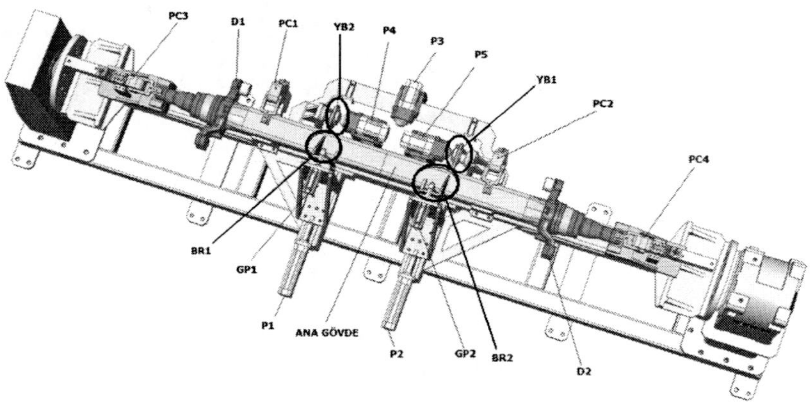

Figue 11: Schematic of Axle Truck

In order to prevent heat based distortions, robots weld the circular welds first (D1,D2). And then BR1,BR2 and YB1,YB2 welds one by one.

HEAVY CONSTRUCTION EQUIPMENT CHASSIS MANUFACTURING

Using twin arc welding process to reduce the takt time. To reach all the welding areas a 2 axis L type positioner and 2 axis (Z , Y) gantry for each robot is used. Total of 3 robots are processing in this project. Loading-unloading of the work parts to the welding fixtures are done by an autonomous sliding system.

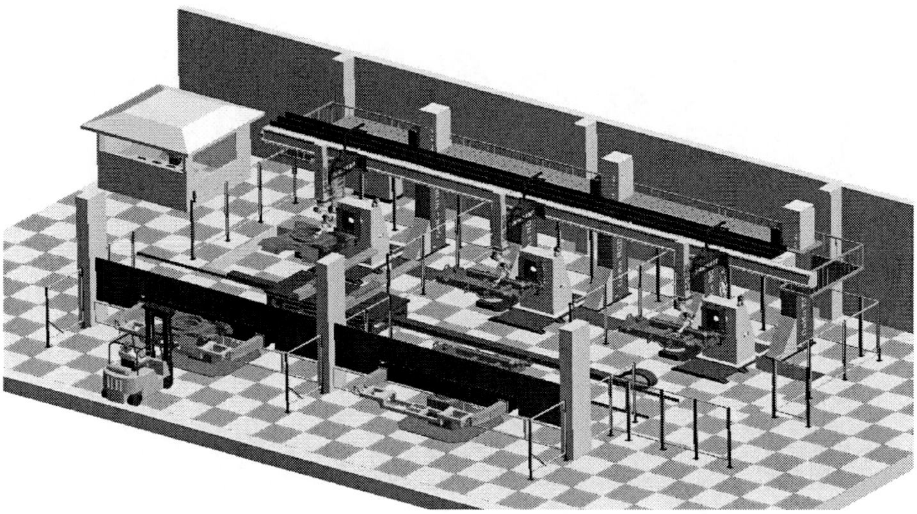

Figure 12: Layout of heavy construction equipment chassis manufacturing

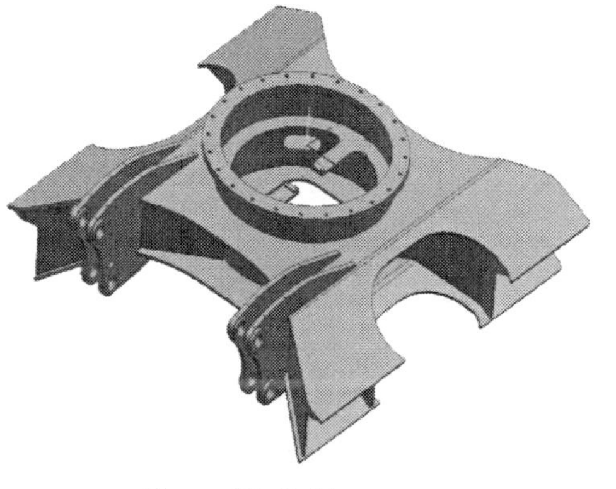

Figure 13: Spider

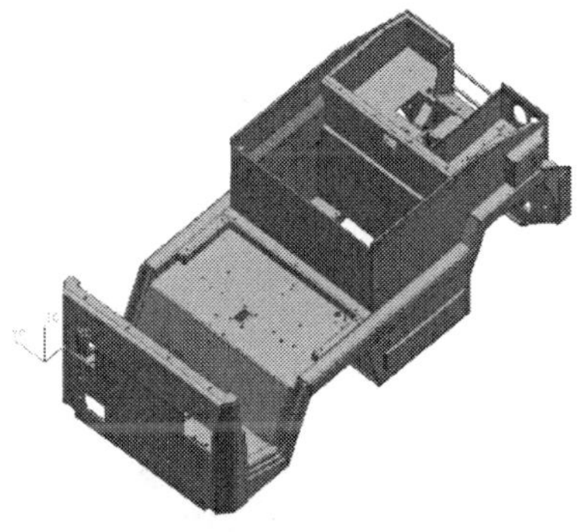

Figure 14: Airport baggage carrier

Figure 15: Lower chassis of construction machine

AGRICULTURAL MACHINE APPLICATION

This system aims to weld the complete Gobley chasiss of the agricultural machine.

Figure 16: Adjustment of touch points and welding parameters

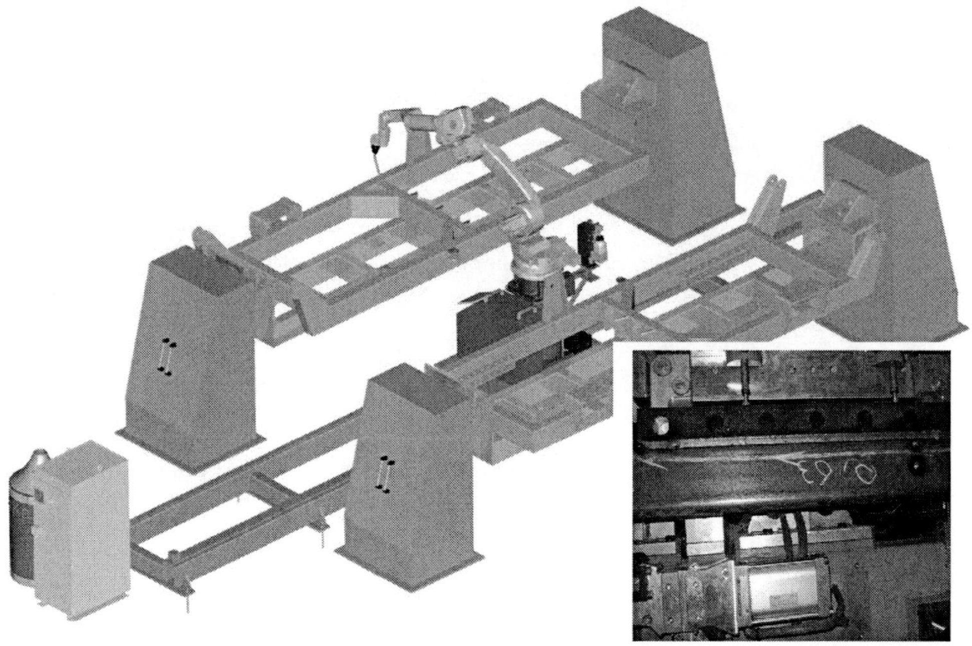

Figure 17: Layout of the complete Gobley chasiss of the agricultural machine and the weld quality

PRESSURE VESSELS MANUFACTURING

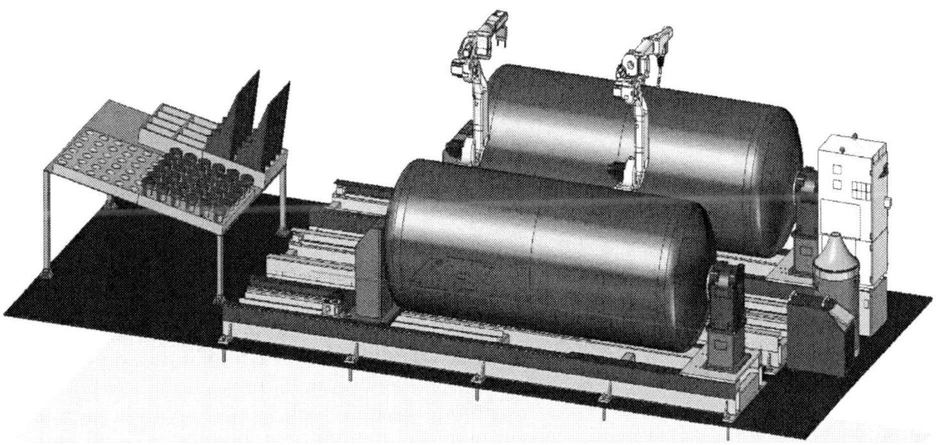

Figure 18: Layout of the stainless steel pressure vessels welding

ORBITAL WELDING SYSTEM

Pipelines are the safest and the most economical method for transmitting large amount of oil and natural gas from remote areas to consuming markets. Ever-increasing weld quality and productivity requirements of the final welds from the pipeline, orbital welding is the first choice for the joining of pipes. Due to the automation, consistent and excellent weld quality, orbital welding is far greater than what can be achieved manually. Orbital welding is an automated arc welding method that moves the arc around a round work piece while the work piece remains stationary. Because of the automatization, the system provides enhanced precision, reliability and time-saving compared to manual techniques.

Figure 19: Application of orbital welding system

Pipelines with different pipe diameters as small as a few millimeters or as large as several centimeters can be welded in this way. High deposition rate with solid and flux cored wires for root, hot, filler and cap passes can be reached by this automatic MIG-MAG orbital welding process. The orbital system consists of pipe truck, power source, pipe band, wire feed and control panels. During the welding procedure, welding parameters such as dwell time, oscillation speed and width, tractor speed can be adjust by welding operator manually also. This method requires good preparation stages of weld groove and pipe alignment.

If weld groove is not opened with a machine, most attention must be paid to smoothness and homogeneity of the weld groove. To obtain a faultless and quality weld seam in welding applications done in accordance with API 1104, pipes needs to be aligned carefully in order to avoid eccentricity. Correct alignment of the pipes is very crucial especially in the spirally welded pipes where ovality is inevitable. Misalignment between the pipe pieces should not be more than 1,5mm.

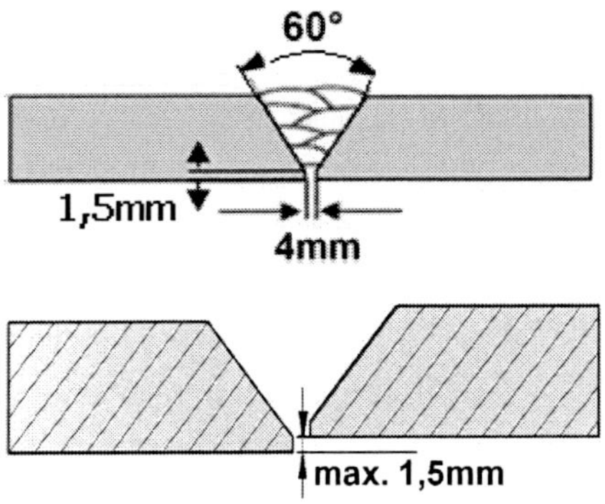

Figure 20: Weld groove preparation of pipes

CONCLUSION

The trend towards automation and application of robots in the welding process are continuing due to the increasing industrialization of huge growing consumer markets, development of new technologies usage of new materials in the industry, the necessary modernizing of existing production facilities and the need to be competitive in a global market. Energy-efficiency and light weight construction are the main challenges of the manufacturing industry. The potential of the use of industrial robots in the Industry will substantially grow when robots will collaborate with workers and when the integration of the robot will become easier.

REFERENCES

1. I.Karabegović, E.Karabegović, S.Pašić and S. Isić, "Worldwide Trend of the Industrial Robot Applications in the Welding Processes" International Journal of Engineering & Technology IJET-IJENS Vol: 12 No: 01.
2. G. Bolmsjö, M. Olsson and P.Cederberg, "Robotic Arc Welding — Trends and Developments for Higher Autonomy" Division of Robotics Department of Mechanical Engineering Lund University, Sweden, 14 November 2001.

3. International Federation of Robotics (IFR) (www.ifr.org)
4. R. Schneider, "Robotics and Automation Can Save Your Factory" President & CEO FANUC Robotics Americas, Inc. Rochester Hills, Michigan.
5. G. Vargın Gök, Ç. Afyon," Kaynak Uygulamalarında Robot Teknolojisi" Kaynak Teknolojisi II. Ulusal Kongresi p: 221-230.
6. Robotics and Automation Can Save Your Factory by Rick Schneider President & CEO FANUC Robotics Americas, Inc. Rochester Hills, Michigan.
7. K. Yu-ichi, "Overwiev of Recent Welding Technology Relating to Pipeline Construction" Transactions of JWRI, Vol.37 (2008), No.1.
8. D. Gençkan and M.Koçak," Orbital Kaynak Teknolojisi ile Doğal Gaz Boru Hattı Kaynaklarının Geliştirilmesi", ITÇ 2012- III. İleri Teknolojiler Çalıştayı 4–6 Ekim 2012.

Thermal weld seam inspection in pipe production lines

V. Schauder, T. Köhler, Halle (Saale)/D, B. Wenzl, Judenburg/AT, M. Prasek, Strongsville/US, and M. Schmitt, Rockton/US

Introduction

In 2007 we began to develop the following thermography system with the goal of finding the position of MAG seams and holes in TIG welding robot application. We knew that only measuring electric parameters (voltage, current) would not allow the detection of all welding failures. We sought new methods to find additional information about the seam.
The idea was to measure the temperatures near the welding point. Only by developing our own thermo line camera (Thermoprofilescanner) could this be achieved. We learnt by doing, we measuring heat signatures and how to interpret them.
Later in 2009 we began to apply this method to tube and pipe welding, and opened a new application field. We began to understand, that we were not only measure the surface temperature. What we do is really the applied passive thermography method. We now have the possibility to look inside the material, to see the penetration.

In this lecture we will explain this NDT method based on passive thermography and this equipment. It is very new and not yet a quality control standard anywhere in the world.

Approaches to Nondestructive inspection of welding quality
 Offline: Inspection in a special inspections station at the end of the production line
 Inline: Inspection during the production, near to the welding point

The inline inspection offers two big advantages:
a)If occurs problems, the mill can stop and will not produce more faulty tubes .
b)Based on the monitoring quality the machine parameters can be optimized continuously by the operator. Typical problems can be prevented.

This method can only be used with inline inspection.

Weldseam inspection in continuous seam tube mill

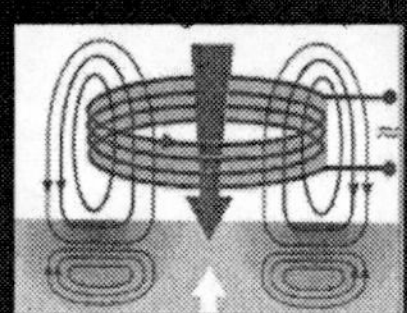

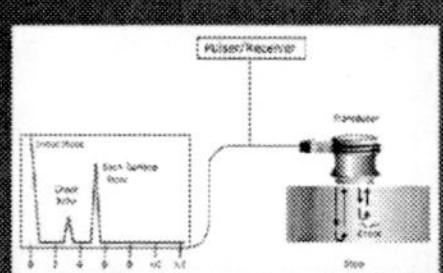

- Most common method of tube production is continuous welding in a mill. Forming the tube from a band and cutting to length after welding. Arc , Laser and HFI welding are used.
- Many international standards require Eddy current (ET) and / or Ultrasonic testing (UT).
- ET is based on the difference in the magnetic flows along the seam. UT works based on the reflection on surfaces or cracks.
- In Arc welding processes this methods based on his physical properties and not sensitive enough for penetration problems in the root, unsymmetrical penetration, root seam forming.
- In HFI welding processes this methods can't find cold welds, one of the most critical problems.

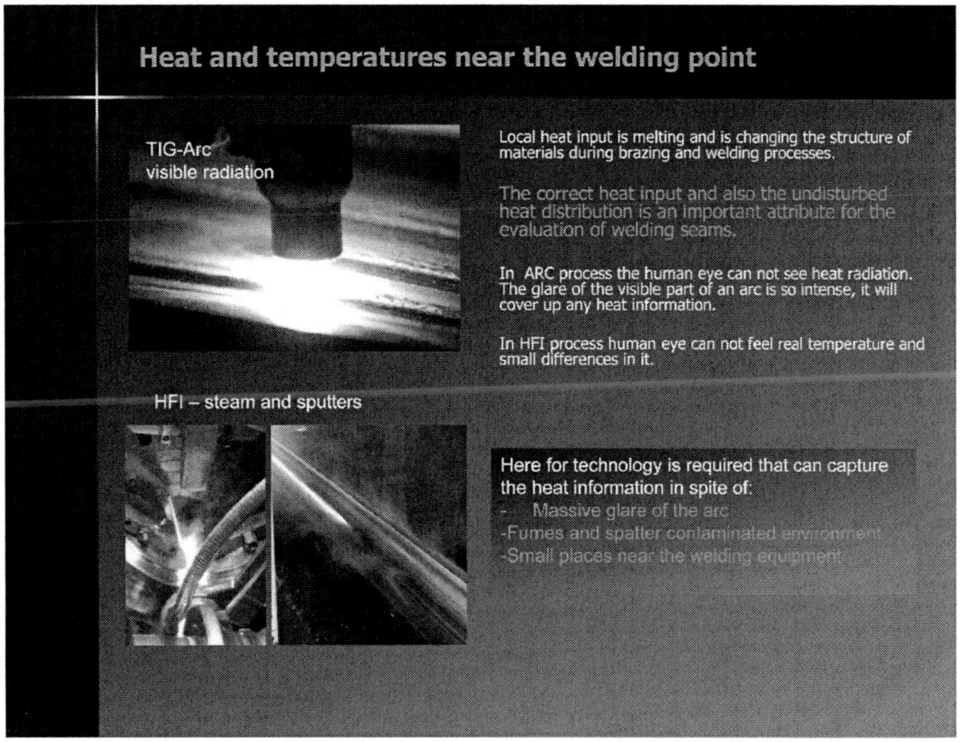
Heat and temperatures near the welding point
TIG-Arc
visible radiation
Local heat input is melting and is changing the structure of materials during brazing and welding processes.
The correct heat input and also the undisturbed heat distribution is an important attribute for the evaluation of welding seams.
In ARC process the human eye can not see heat radiation. The glare of the visible part of an arc is so intense, it will cover up any heat information.
In HFI process human eye can not feel real temperature and small differences in it.
HFI – steam and sputters
Here for technology is required that can capture the heat information in spite of:
- Massive glare of the arc
-Fumes and spatter contaminated environment
-Small places near the welding equipment

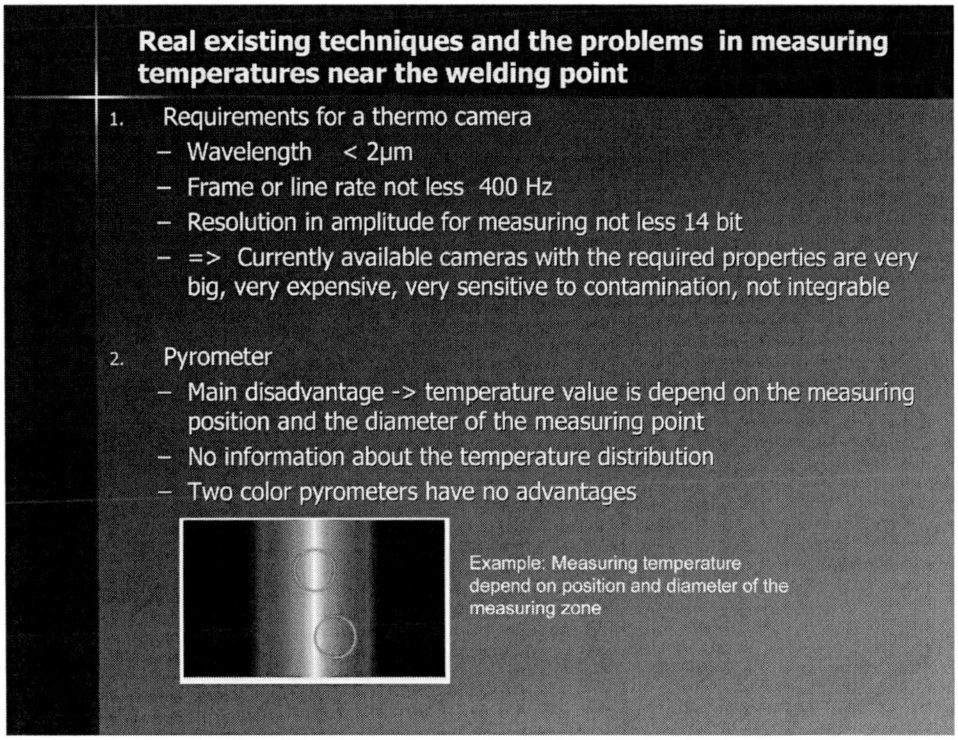
Real existing techniques and the problems in measuring temperatures near the welding point
1. Requirements for a thermo camera
– Wavelength < 2μm
– Frame or line rate not less 400 Hz
– Resolution in amplitude for measuring not less 14 bit
– => Currently available cameras with the required properties are very big, very expensive, very sensitive to contamination, not integrable
2. Pyrometer
– Main disadvantage -> temperature value is depend on the measuring position and the diameter of the measuring point
– No information about the temperature distribution
– Two color pyrometers have no advantages
Example: Measuring temperature depend on position and diameter of the measuring zone

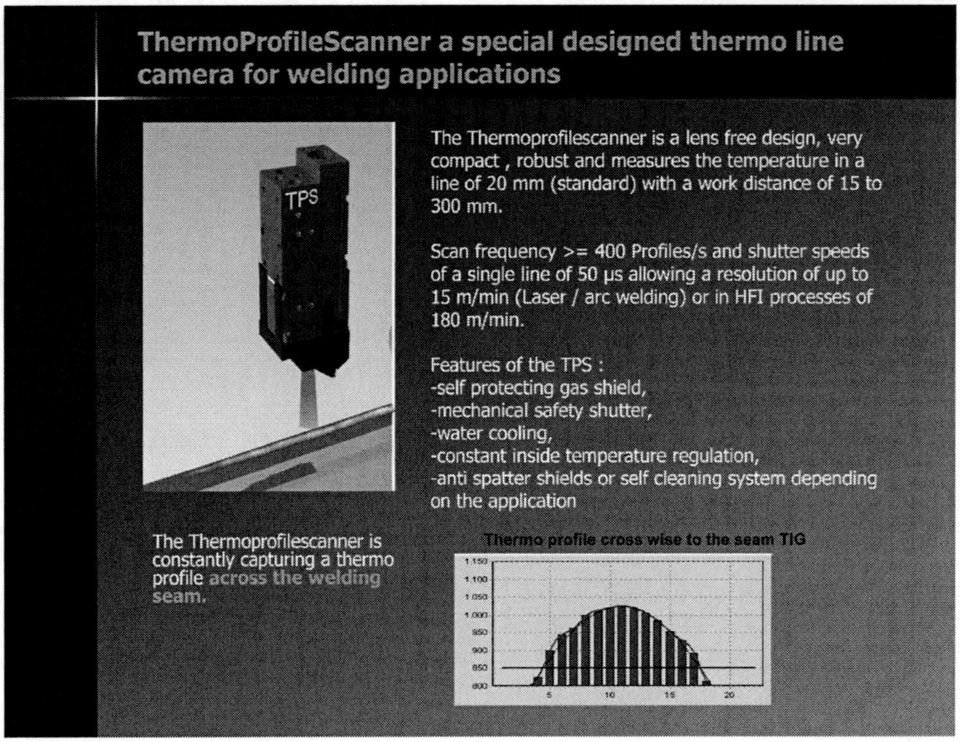
ThermoProfileScanner a special designed thermo line camera for welding applications
TPS
The Thermoprofilescanner is a lens free design, very compact , robust and measures the temperature in a line of 20 mm (standard) with a work distance of 15 to 300 mm.
Scan frequency >= 400 Profiles/s and shutter speeds of a single line of 50 µs allowing a resolution of up to 15 m/min (Laser / arc welding) or in HFI processes of 180 m/min.
Features of the TPS :
-self protecting gas shield,
-mechanical safety shutter,
-water cooling,
-constant inside temperature regulation,
-anti spatter shields or self cleaning system depending on the application
The Thermoprofilescanner is constantly capturing a thermo profile across the welding seam.
Thermo profile cross wise to the seam TIG
1.150
1.100
1.050
1.000
950
900
850
800
5
10
15
20

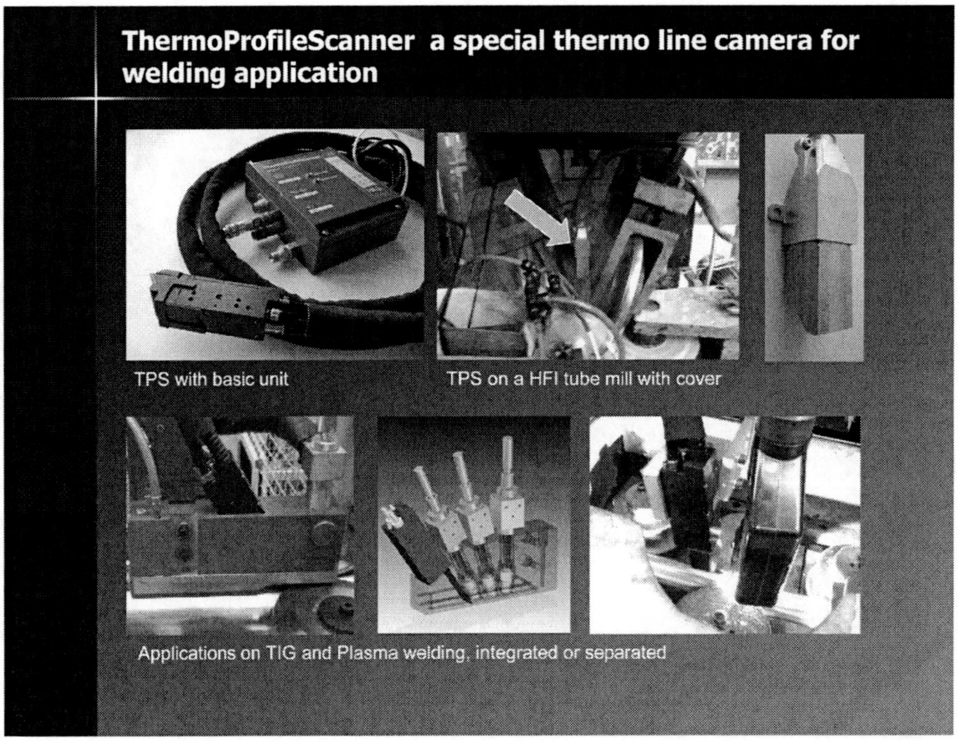
ThermoProfileScanner a special thermo line camera for welding application
TPS with basic unit
TPS on a HFI tube mill with cover
Applications on TIG and Plasma welding, integrated or separated

Design and method of operation

1. The Thermoprofilescanner is capturing the temperature over the welding joint and is sending the data to the WeldQAS-device.
2. The WeldQAS-device is calculating for each line the attributes of the profile.

Calculated from the heat signature for every profile online:

- max temperature
- width of temperature zone
- symmetry
- profile position
- form differences

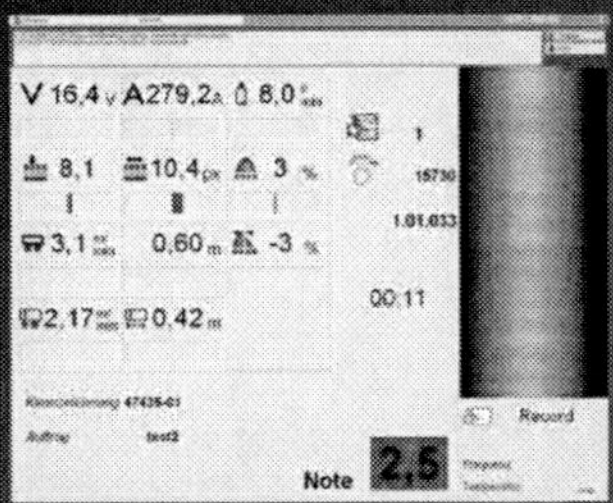

3. The heat images are displayed simultaneous visually by the WeldQAS, stored and compared with programmed set values.
4. By recognizing violations of limit values the unit detecting welding irregularities and their position on the tube.

Design and method of operation

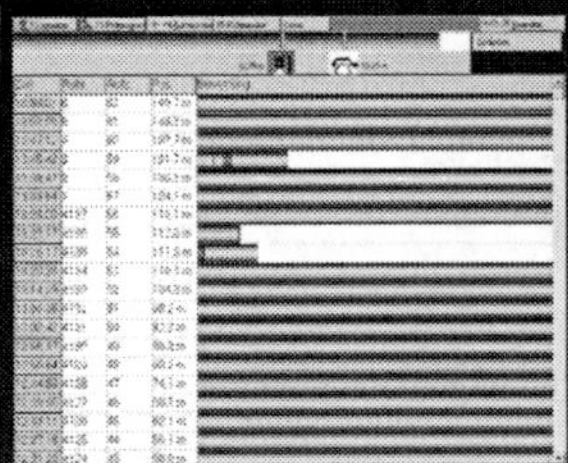

Monitoring actual state of tubes and faults

5. The error signal is generated immediately or can be buffered to mark the defective with a marking spray unit.
6. The WeldQAS-device is storing all data pertaining to the pipes, which will be numbered and can be synchronized with a saw signal.
7. The data are stored in a data base and are displayed in a pipe monitor program.

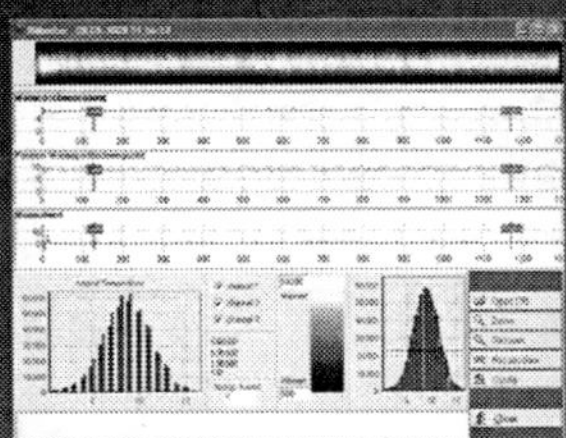

Head signature of a real 6 m tube

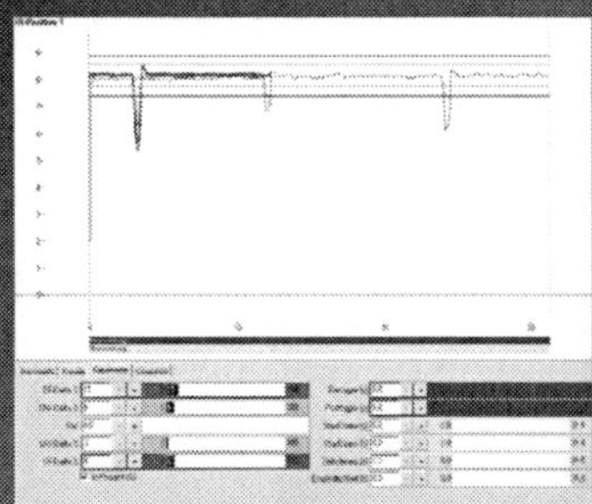

Program for setting the warning and error values based on the real measuring results

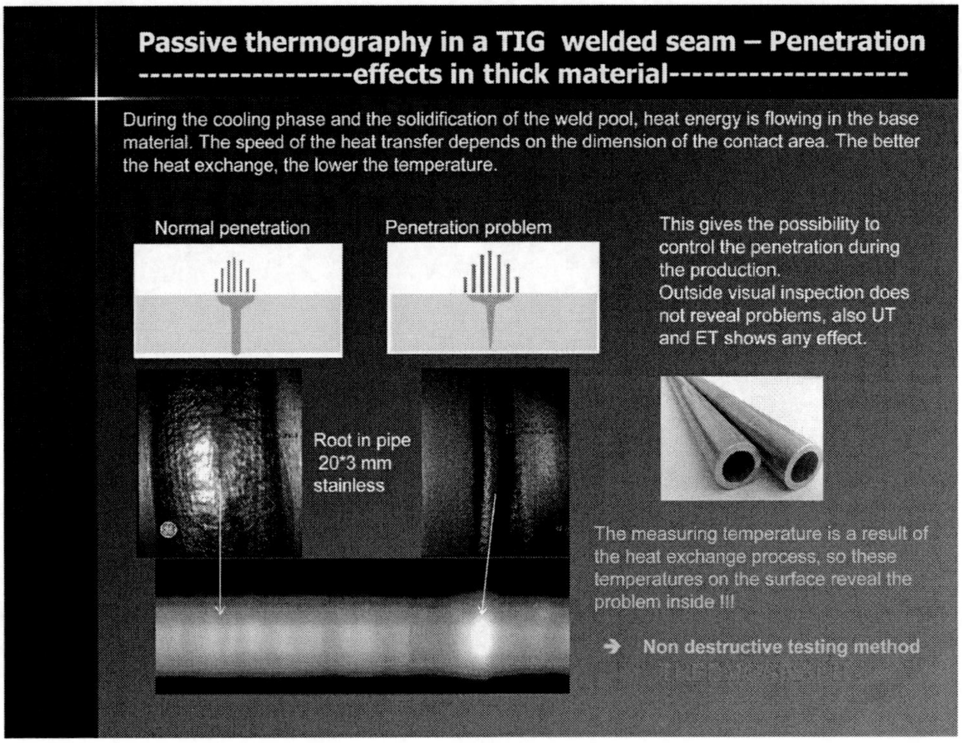
Passive thermography in a TIG welded seam – Penetration
-------------------effects in thick material---------------------
During the cooling phase and the solidification of the weld pool, heat energy is flowing in the base material. The speed of the heat transfer depends on the dimension of the contact area. The better the heat exchange, the lower the temperature.
Normal penetration
Penetration problem
This gives the possibility to control the penetration during the production.
Outside visual inspection does not reveal problems, also UT and ET shows any effect.
Root in pipe
20*3 mm
stainless
The measuring temperature is a result of the heat exchange process, so these temperatures on the surface reveal the problem inside !!!
➔ Non destructive testing method

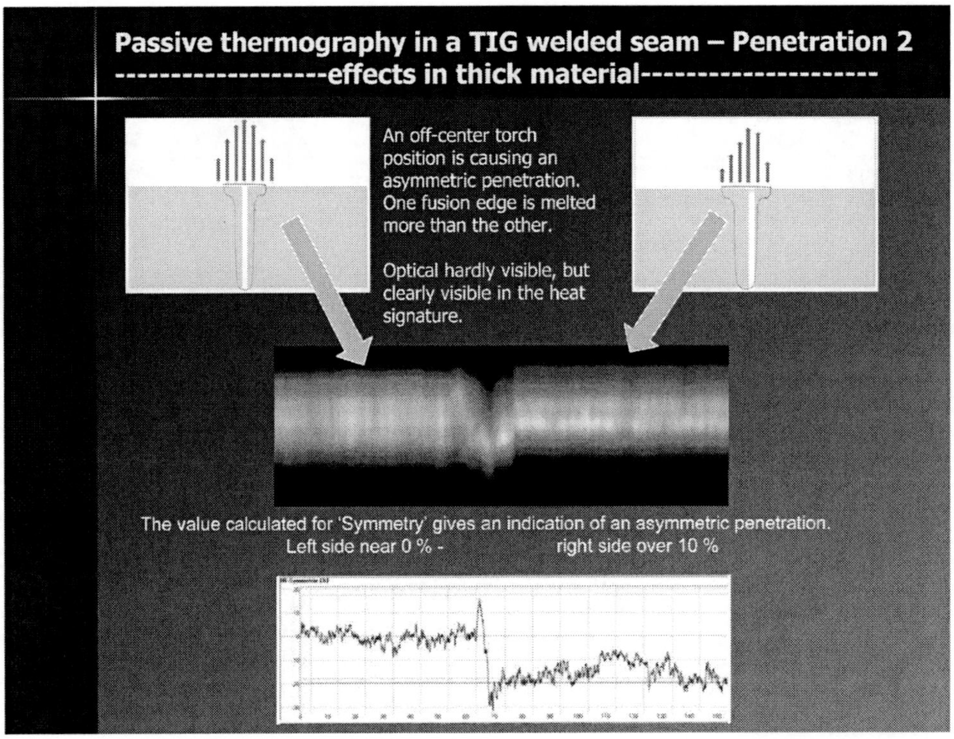
Passive thermography in a TIG welded seam – Penetration 2
-------------------effects in thick material---------------------
An off-center torch position is causing an asymmetric penetration. One fusion edge is melted more than the other.
Optical hardly visible, but clearly visible in the heat signature.
The value calculated for 'Symmetry' gives an indication of an asymmetric penetration.
Left side near 0 % -
right side over 10 %

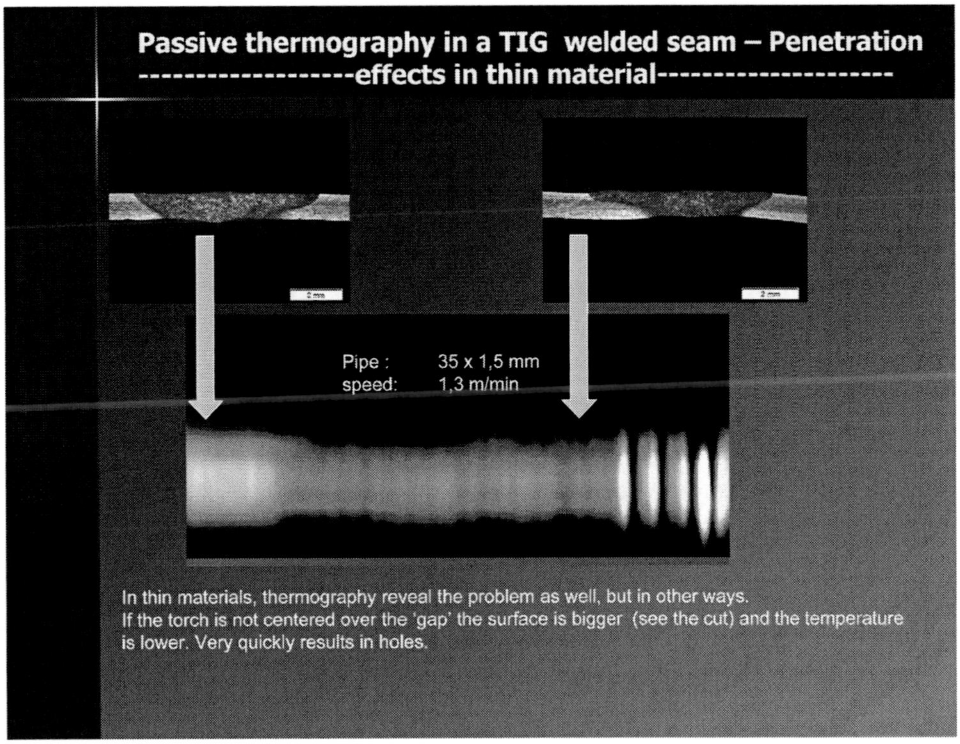
Passive thermography in a TIG welded seam – Penetration
------------------effects in thin material--------------------
Pipe : 35 x 1,5 mm
speed: 1,3 m/min
In thin materials, thermography reveal the problem as well, but in other ways.
If the torch is not centered over the 'gap' the surface is bigger (see the cut) and the temperature is lower. Very quickly results in holes.

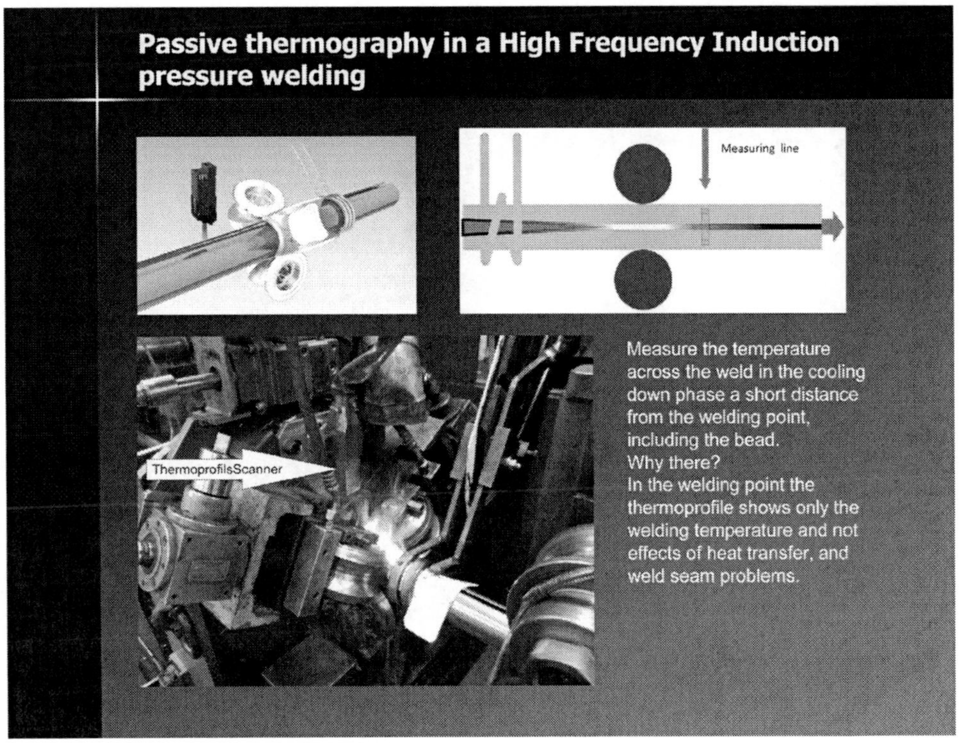
Passive thermography in a High Frequency Induction pressure welding
Measuring line
ThermoprofilsScanner
Measure the temperature across the weld in the cooling down phase a short distance from the welding point, including the bead.
Why there?
In the welding point the thermoprofile shows only the welding temperature and not effects of heat transfer, and weld seam problems.

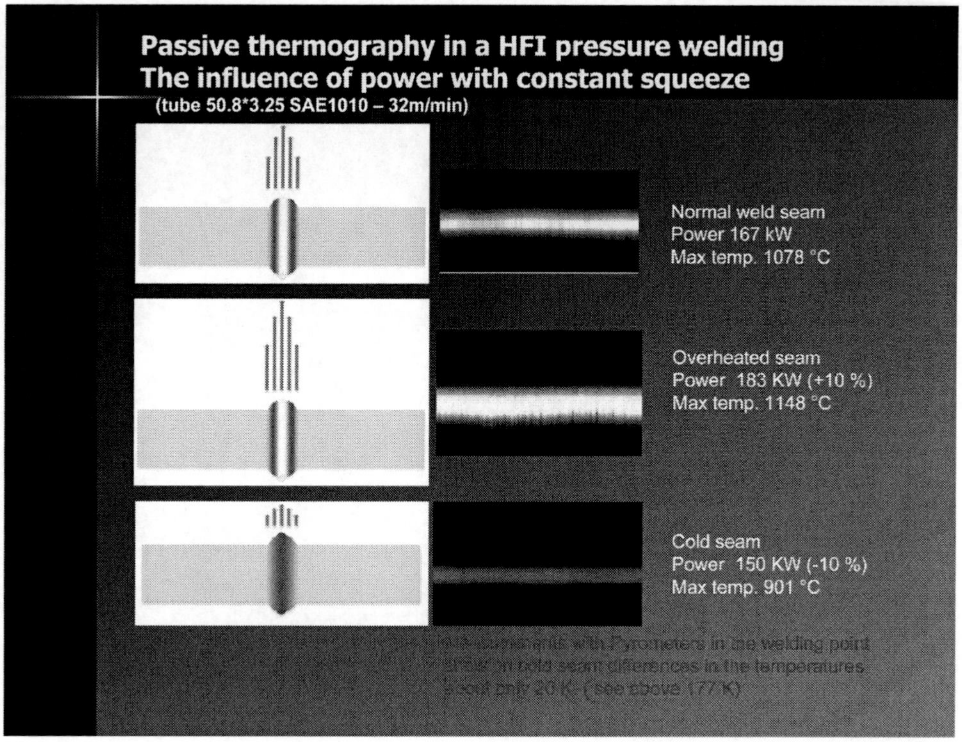
Passive thermography in a HFI pressure welding
The influence of power with constant squeeze
(tube 50.8*3.25 SAE1010 – 32m/min)
Normal weld seam
Power 167 kW
Max temp. 1078 °C
Overheated seam
Power 183 KW (+10 %)
Max temp. 1148 °C
Cold seam
Power 150 KW (-10 %)
Max temp. 901 °C
with Pyrometers in the welding point
differences in the temperatures
(see above 177 K)

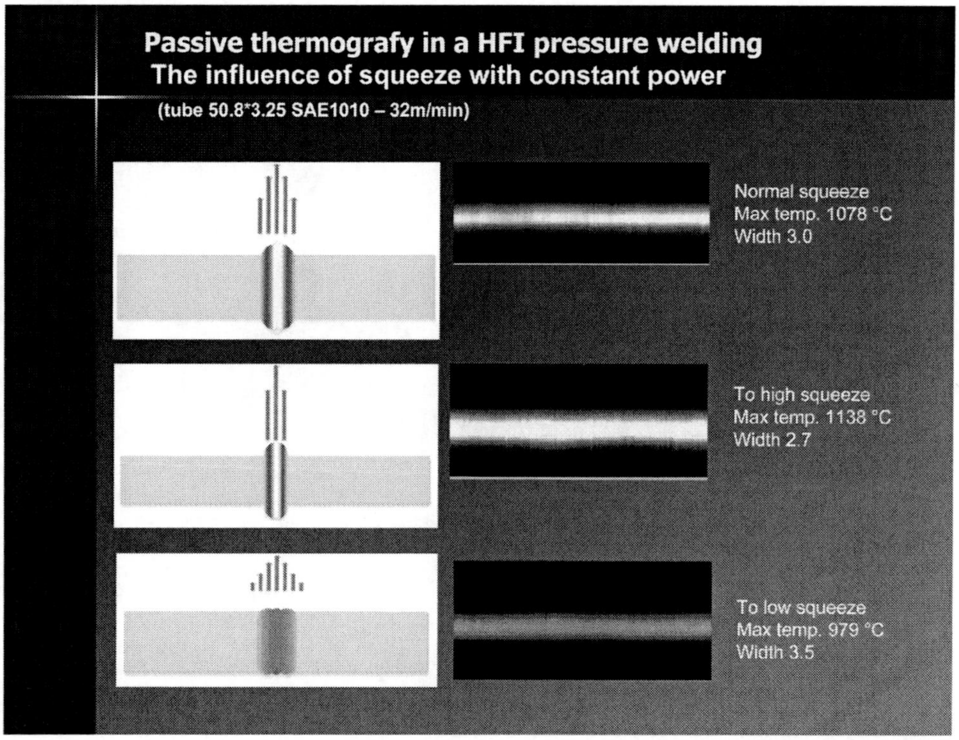
Passive thermografy in a HFI pressure welding
The influence of squeeze with constant power
(tube 50.8*3.25 SAE1010 – 32m/min)
Normal squeeze
Max temp. 1078 °C
Width 3.0
To high squeeze
Max temp. 1138 °C
Width 2.7
To low squeeze
Max temp. 979 °C
Width 3.5

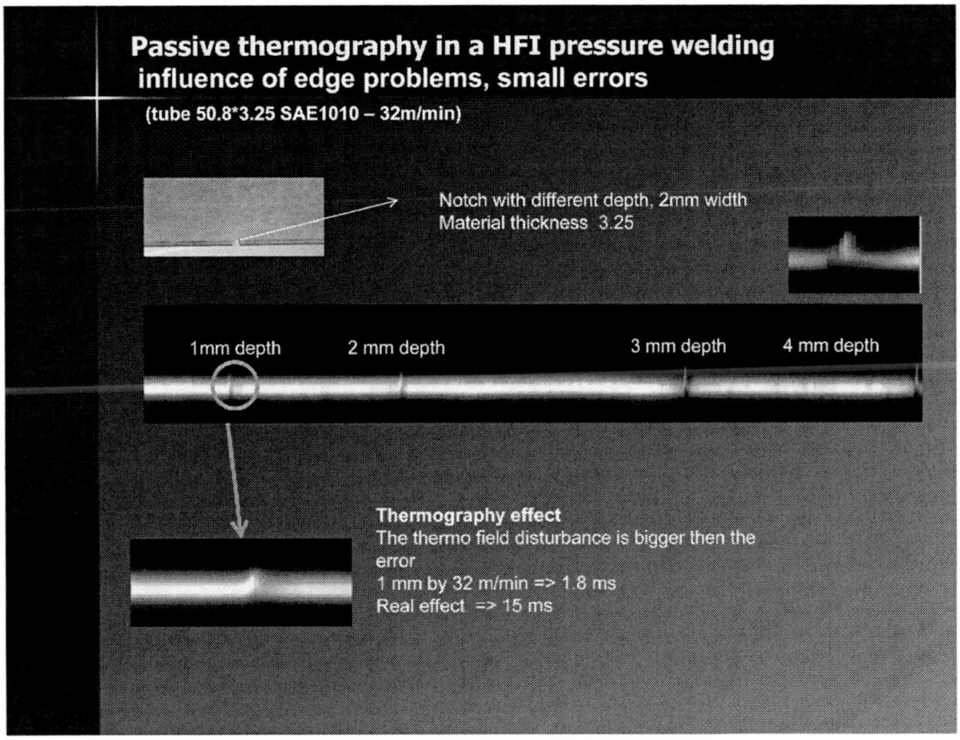
Passive thermography in a HFI pressure welding
influence of edge problems, small errors
(tube 50.8*3.25 SAE1010 – 32m/min)
Notch with different depth, 2mm width
Material thickness 3.25
1mm depth
2 mm depth
3 mm depth
4 mm depth
Thermography effect
The thermo field disturbance is bigger then the error
1 mm by 32 m/min => 1.8 ms
Real effect => 15 ms

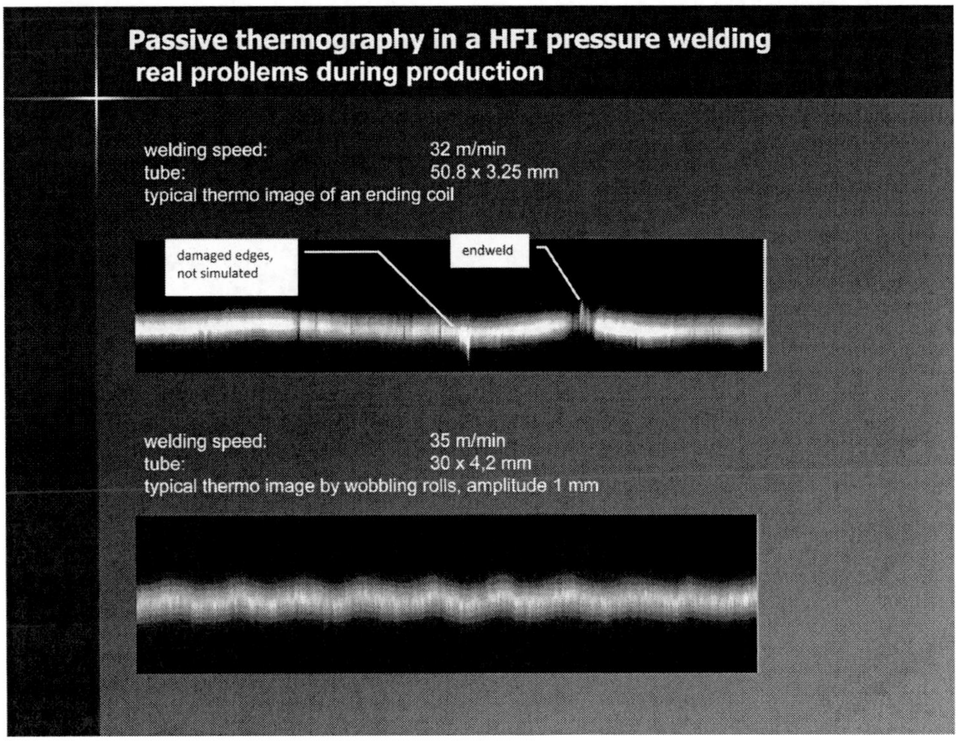
Passive thermography in a HFI pressure welding
real problems during production
welding speed: 32 m/min
tube: 50.8 x 3.25 mm
typical thermo image of an ending coil
damaged edges, not simulated
endweld
welding speed: 35 m/min
tube: 30 x 4,2 mm
typical thermo image by wobbling rolls, amplitude 1 mm

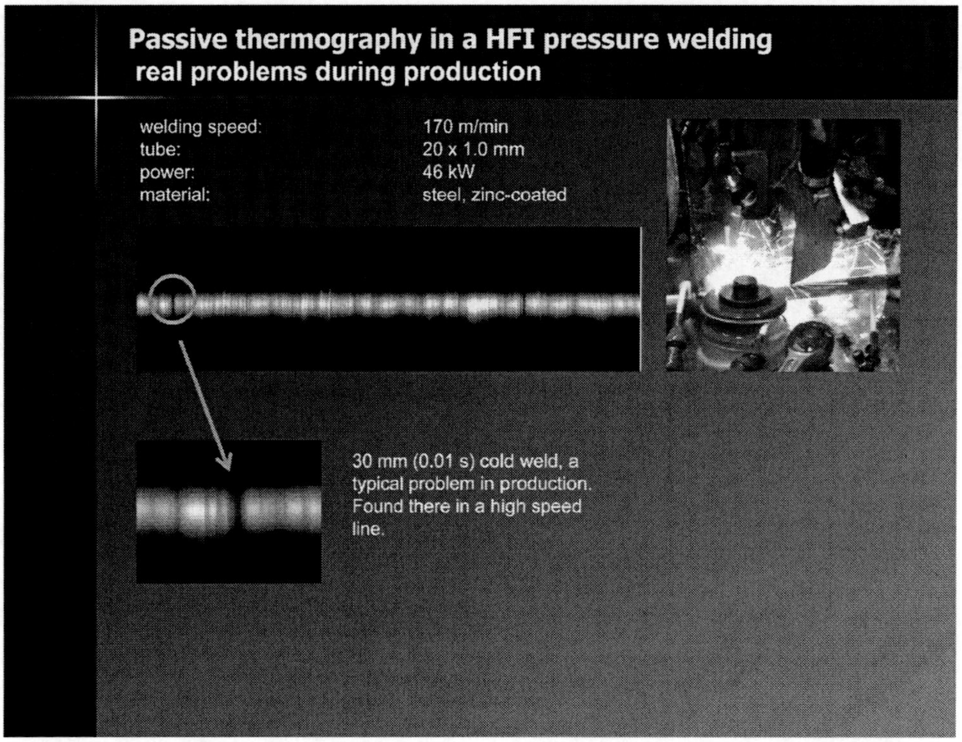
Passive thermography in a HFI pressure welding
real problems during production
welding speed: 170 m/min
tube: 20 x 1.0 mm
power: 46 kW
material: steel, zinc-coated
30 mm (0.01 s) cold weld, a typical problem in production. Found there in a high speed line.

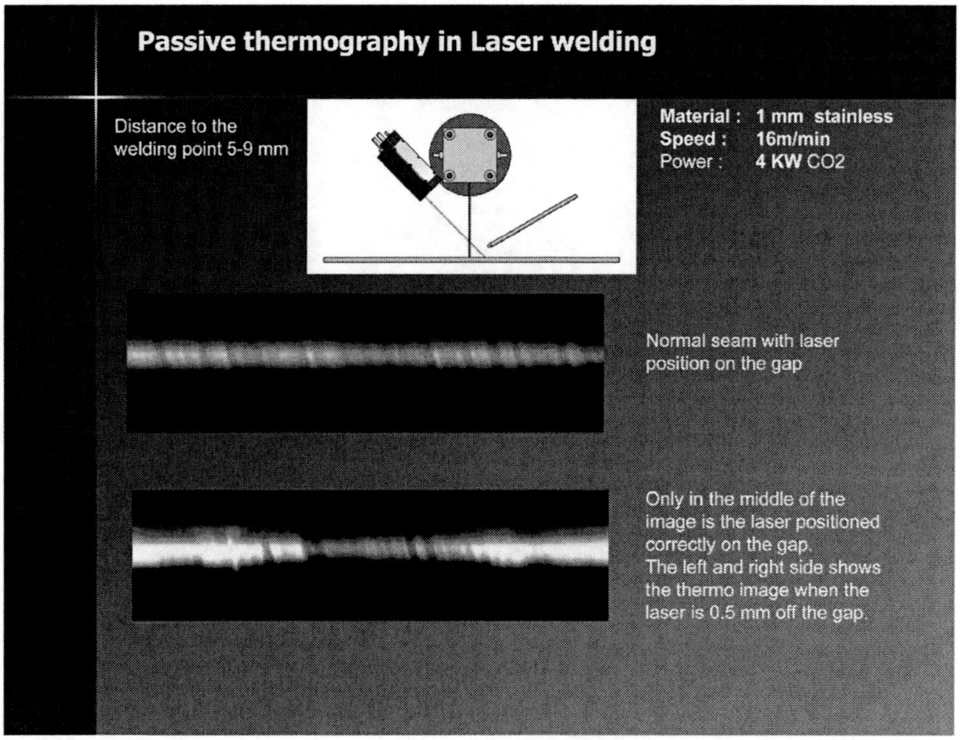
Passive thermography in Laser welding
Distance to the welding point 5-9 mm
Material : 1 mm stainless
Speed : 16m/min
Power : 4 KW CO2
Normal seam with laser position on the gap
Only in the middle of the image is the laser positioned correctly on the gap.
The left and right side shows the thermo image when the laser is 0.5 mm off the gap.

conclusion

- The development of passive thermography used near the welding point created the possibility to measure temperature profiles with high speed and high accuracy. To achieve that a special thermo line camera was created. The Thermoprofilescanner .
- This method is based on differences in temperatures, which occur trough differences in the heat distribution in the down cooling phase of the welding seam. This allows the possibility to see effects inside the material.
- This inline NDT process can be used in mills using TIG, Plasma, HFI and Laser welding and shows better results then eddy current and ultrasonic testing.
- This Thermography non-destructive testing method also works as a monitoring system and shows result parameters of the seam and a running thermo image. This allows tuning the machine parameters to generate continuous high quality seam.
- Over 60 applications around the world in the last three years in tube and pipe welding show the potential of this method.
- Thanks to all our costumers, that helps with theire patience and support to improve this system.

Advanced arc processes meet customer needs for automated steel welding

G. Wilhelm, B. Jaeschke, Auenwald/D, M. Stacey, B. Stacey, Cannock/UK, and M. Weglowski, Warsaw/PL

Structure

- Description of the case study
  → Ambience and boundary conditions
  → Weld task
  → Objectives
- Solution
  → SpeedRoute process
  → Weld results
- Further beneficial processes
  → SpeedUp process
  → SpeedPulse process

Ambience conditions

Ambience conditions

Ambience conditions

Weld task

Parent material: L 485 MB EN 10208-2 (similar to TM X 70)

C	Si	Mn	P	S	Ti	V	Nb
max.	max.	max.	max.	max.	max.	max.	max.
0,16	0,45	1,70	0,025	0,020	0,06	0,10	0,06

Tube: Outside diameter: 1016 mm

Welding process: GMAW

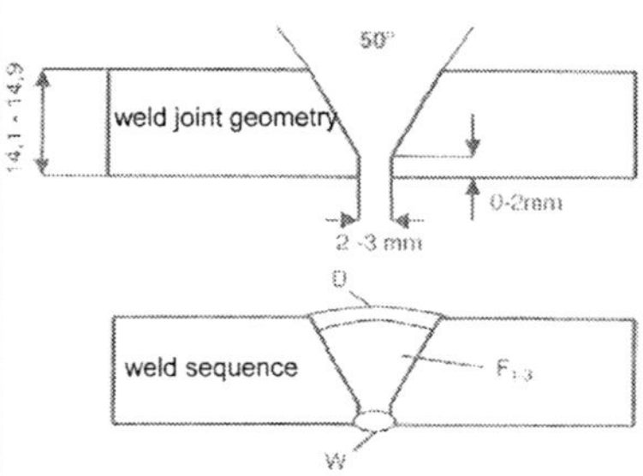

Wire electrode and welding position:

W: G 55 4 C Mn3Ni1 Mo, EN ISO 12534
(Böhler NiMo 1 IG)
Welding position: PJ

F, D: T 46 3 P M 1 H5, EN ISO 17632-A
(Oerlikon Fluxofil 14 HD)
Welding Position PH

Shielding Gas:

W: C1, EN ISO 14175

F, D: M 21; EN ISO 14175

Preheating temperature: 80°C to 200°C

Weld task

Weld problems:

The tubes must be centred and the gap has to be adjusted.
→ local damages of the V-joint preparation
→ local root face differences
→ local gap differences
→ partial lack of root penetration

Top view correct V-joint preparation

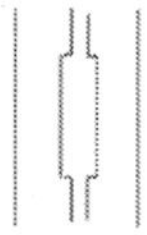

Top view damaged V-joint preparation

Objectives:

One parameter setting per layer for all positions:

→ Welding speed
→ Wire speed
→ Voltage
→ Oscillation width
→ Oscillation speed
→ 2 power sources shall operate at the same time without interacting

High process capability:

→ Reproducable and well shaped excess penetration independent of root face damages
→ No suck back of the root at the overhead position
→ No excessive weld reinforcement of the fianl pass at the overhead position
→ Able to deal with difficult ambience conditions
→ etc.

Good quality:

→ Test standards: DGRL/PED/EC, AD 2000 HP 5/2, EN ISO 15614-1

Solution:

SpeedRoot process:

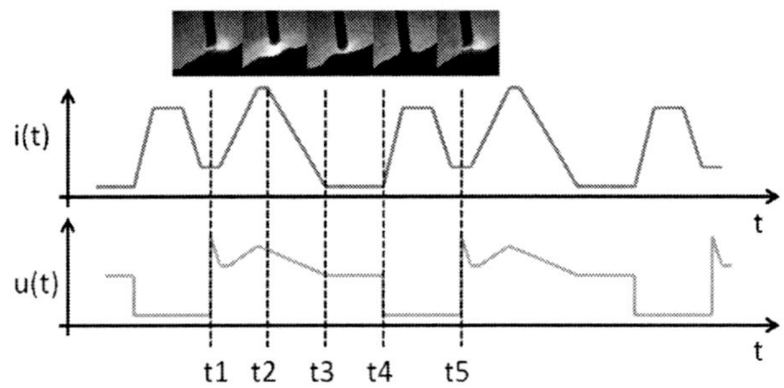

Solution:

At t1 the previous short circuit phase ends while at t4 a new short circuit phase is going to establish.

At t5 the short circuit phase between t4-t5 ends. The duration between t1-t5 is a complete cycle, which is not set as fix. It depends on instantanious process changes and is therefore a result of the process itself.

Current and Voltage are set values realised by the control mode for each defined cycle sector as well as the control mode (constant current or constant voltage).

As all these settings are part of programmed synergy lines the machine is very easy to use. The user just has to chose the wire speed. Small process-adjustments can be done easily

Benefits are a very low spatter generation due to the low current at the moment when the jumper brakes through and a powerful arc because of the high current in the arc burning phase. A boost impulse can increase the penetration.

Results:

external view of the tube

Results:

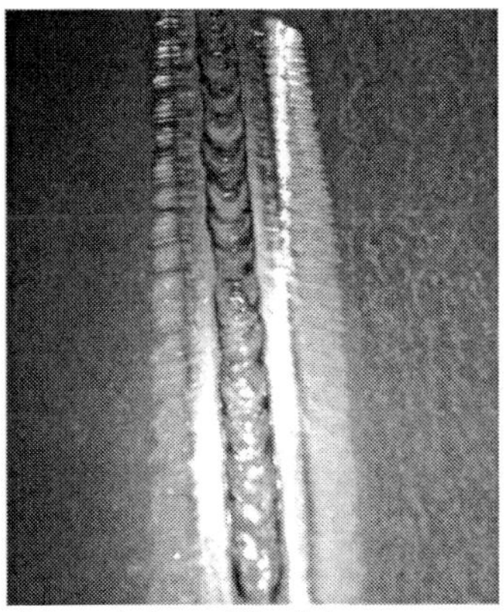

internal view of the tube

Results:

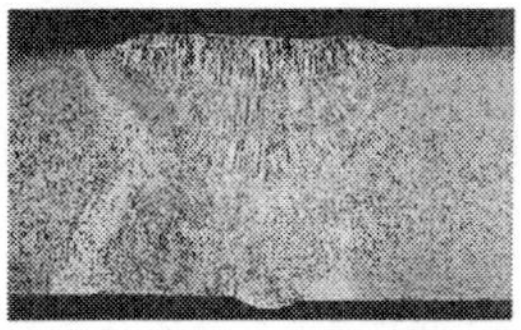

Position 0°

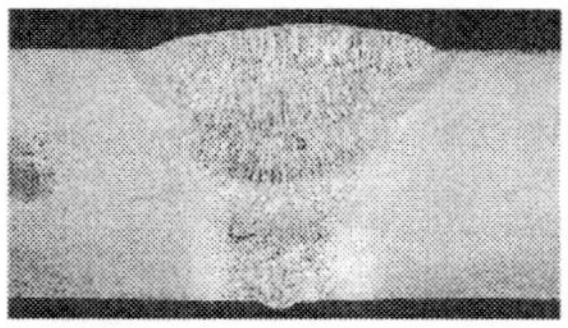

Postion 90°

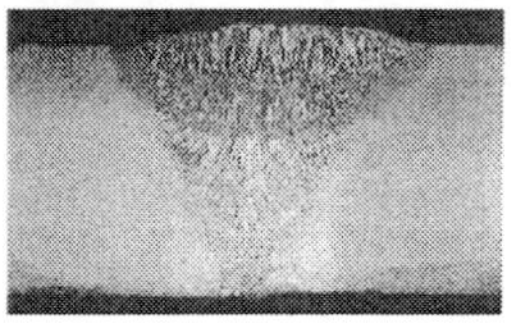

Position 180°

Results:

Impact Test (DIN EN ISO 9016), Specimen Type: ISO-V-KV-300

Abmessung Dimension mm	Prüftemp. Test Temp. °C	Kerblage Notch Location	Mittelwert avarage J / cm²	Ergebnis Result
			>27	
8 x 9,8	- 21	VWT	210	e / acceptable
8 x 9,8	- 21	VHT	114	e / acceptable
8 x 8	- 21	VWT	116	e / acceptable
8 x 8	- 21	VHT	206	e / acceptable

Hardness Test DIN EN ISO 9015-1

- Weld metal deposit
- Heat effected zone
- Parent material

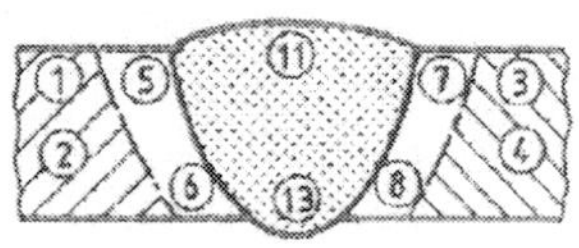

All measurements < 260 HV10
Requirement: max. 380 HV10 without post weld heat treatment

Further beneficial processes:

SpeedUp: A process developed for welding in difficult positions

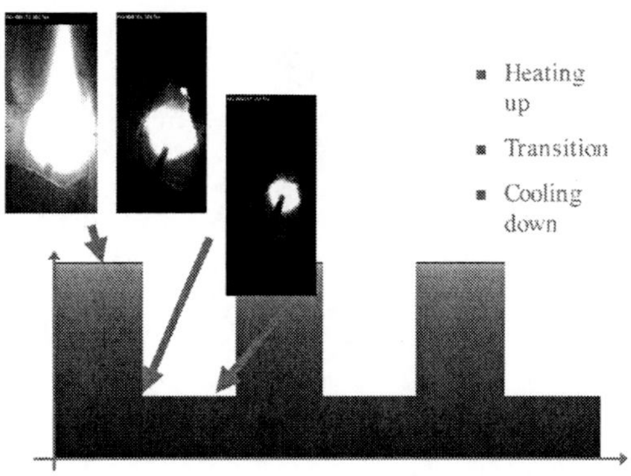

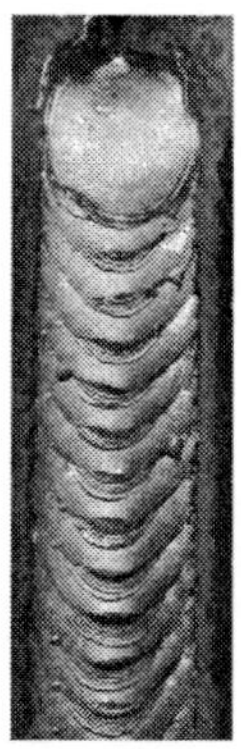

Further beneficial processes:

SpeedUp: A process developed for welding in difficult positions

SpeedUp works by combinig 2 different energy levels periodically.

→ SpeedPulse energy level to insure a good root penetration

→ Modified short arc energy level to enable position welding.

→ Linear vertical up welding is possible (no weaving, triangle welding necessary)

Further beneficial processes:

SpeedPulse: A process developed for fast welding with a high deposition rate

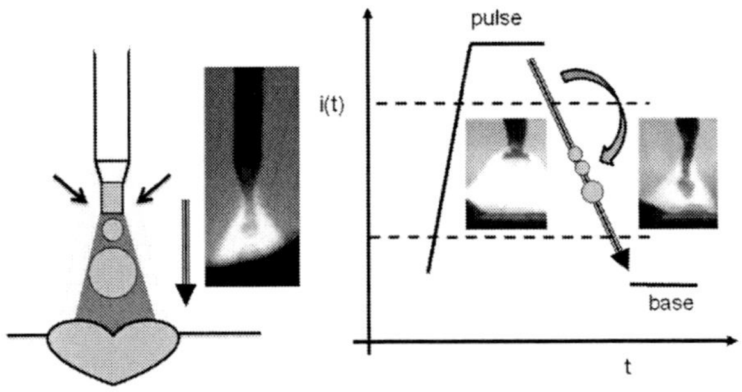

Further beneficial processes:

SpeedPulse: A process developed for fast welding with a high deposition rate

Advantages

→ high deposition rate

→ fast welding possible

→ focused arc

→ good penetration

Improvements in pipeline welding

S. Rajagopalan, Houston/US

Throughout the pipeline industry, and particularly in offshore and spool base production, welding requirements are shifting towards higher quality, greater accuracy and more precision. At the same time, everyone is concerned with pushing productivity levels beyond the old ideas of normal. Only a new technology can address both of these needs and developing an engineering solution for this trend in the industry is a complex process. As the world leader in pipeline welding, CRC-Evans Automatic Welding is in a unique position to achieve a breakthrough by introducing a new category of welding process and equipments that uses this advanced laser technology. This new category of laser inspection and welding systems includes the bevel inspection V Bevel, fully automatic purge control system V Purge, root inspection V Root, external welding machine V Weld and cap inspection V Cap. All have laser vision technology incorporated with CRCs state of the art technology which makes it the first of its kind to be ever used in this field.

Robotic SAW welding on narrow gap profiles with the use of laser scanners

A. Gadalov, Narva/EE

Straight to business

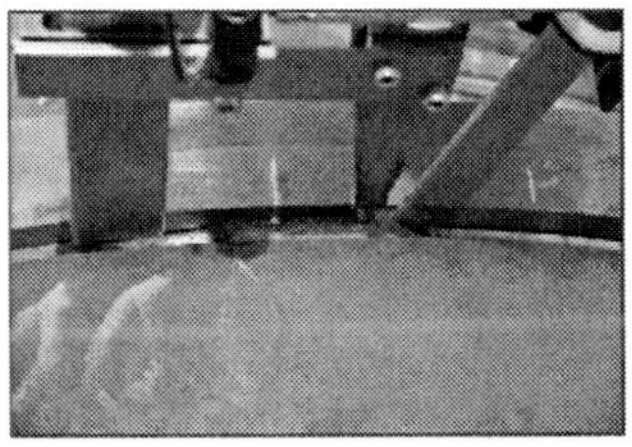

I am pleased to introduce you to some of our latest developments in the field of automatic submerged arc welding (SAW).

By creating robot-based solutions we have achieved significant increase in welding quality and productivity.
Our solutions can be used in a variety of industries where multipass welding of circumferencial, longitudinal and saddle-shaped welds are applicable.

Company's profile

ASG Robotic Ltd is a modern European machine-building company. The company was founded in 2010 as a subsidiary of the Russian reputable firm Kvadrat SG Ltd which since 1985 has been involved in research, development, manufacturing and selling of automated turnkey solutions for such industries as: automative, shipbuilding, piping, chemical and food. In total more than 800 different projects were successfully implemented. Both companies are family operated by Sergey Gadalov as its CEO.

Some of our customers are: Ford Motor Company, Nissan, Caterpillar, Toyota, Volkswagen, Scania, Magna, GTS Technology, Durr, Schneider Electric, Electrolux, Philip Morris, SMC-Pneumatik, Gillette, Wrigley, JTI, Henkel, and many others.

Robots vs. traditional welding manipulators

Why do we use industrial robots?

1. The robot is a complete system, based on which we can build whatever we need.
2. Tens of thousands of robots are made each year, which speaks about their reliability.
3. A robot is a programmable manipulator which can perform a variety of manually trained or mathematically calculated trajectories.
4. Robot Controller allows you to control the entire periphery of the including welding parameters which ensures the stability and quality.
5. User-friendly interface allows you to keep track of each sensor or parameter in the whole system.

All that we want to show you on the following examples of our success stories.

Success stories. Fully welded ball valves

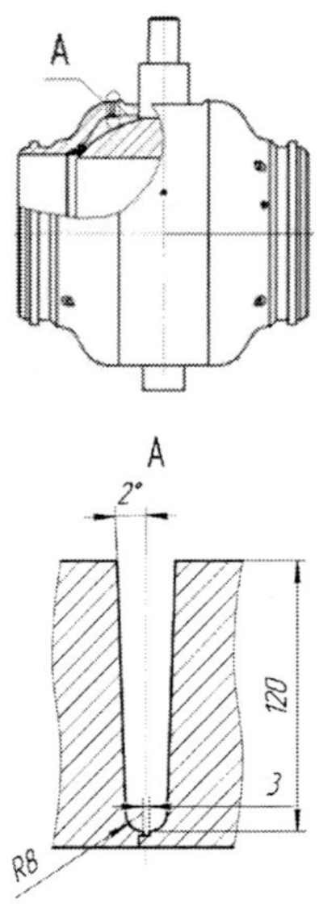

General information.
These fully welded by two circumferential welds ball valves are normally used in oil and gas transport systems.

Technical characteristics
Narrow gap welding profile.
Nominal size: DN 300-1400 mm
Weld profile depth range: 25-120 mm
Maximum weight: 30000 kg
Maximum diameter of welds: 2300 mm
Since there are polymers inside a cooling system is required.

Customer's technical requirements.
Parallel and independent welding of two welds
Compensation of deformation and abnormalities of the gap
Forced cooling and temperature control in the welding area.
Automatic feeding and recycling of flux.

Success stories. Fully welded ball valves

According to our customer's requirements, we built two systems in order to cover a vast range of sizes. Both systems are identical, based on the two-armed robot FANUC R-710iC and single-axis positioner.
Robot's arm positions the welding torches while the positioner rotates the part.

Success stories. Fully welded ball valves

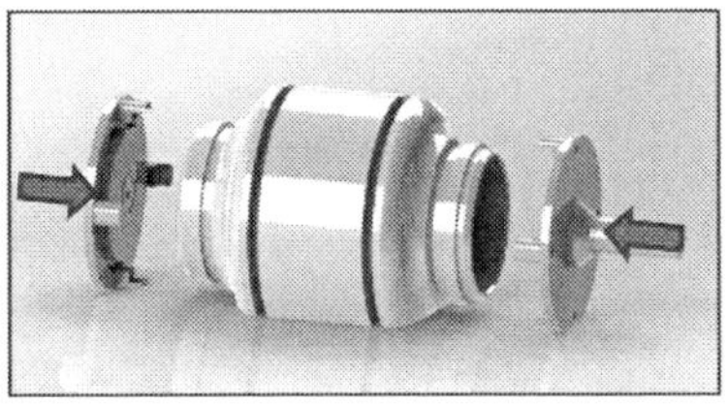

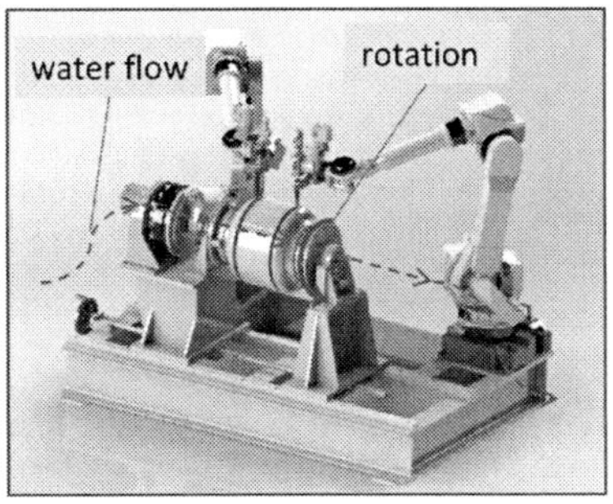

The positioning and cooling.
The valve covers are installed on the valve to make sure the cooling system is leak proof.

After the valve has been placed on the bearings, the gap search process is started so that the robotic system can automatically start welding the root pass and all the subsequent passes with the help of the laser positioning system.

Once the root pass has been weld, the cooling system starts working and monitoring the temperature of the work pieces to avoid overheating the polymeric seals.

Owing to this and many other implemented automation solutions, the efficiency, quality and flexibility of the process has been increased.

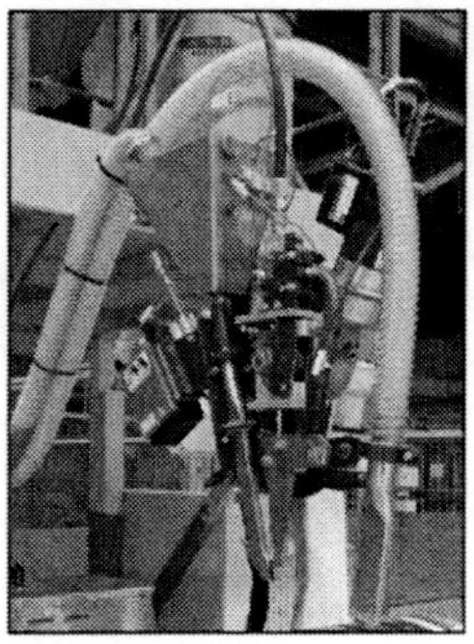

The welding head.

A special welding head has been developed by us for this task. It has the right and left-hand version. Compact and easy to maintain. It has everything you need: a sensor for collision detection, wire sensor, flux sensor, push-inlet valve for flux feed rate control , protection against welding wire stick, wire straightener, laser scanner system to monitor the gap. The thickness of the welding gun is just 10 mm. We also developed a flux recirculation system, which allows to significantly reduce the load on the robot and ensure movement flexibility.

Success stories. Fully welded ball valves

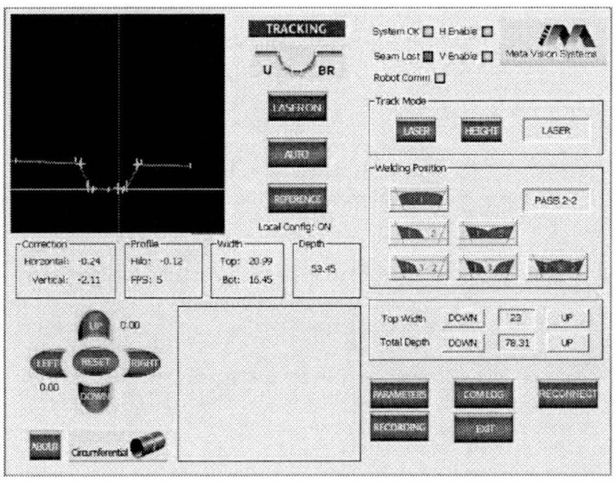

The tracking system.

VistaWeld is a tracking system that searches, finds and tracks weld joints and weld seams using optical triangulation. It provides a reliable non-contact method for controlling the position of welding torches. The system is connected to the robot's controller via Ethernet.

Vista Weld was successfully integrated with the robot controller by a common protocol, wich let the robot have access to the parameters and states of the laser system without having to require the user to change them manually on Vista weld's panel.

Operating system.
The high computing power of the robot controller enables you to control all the devices of the system, including the laser tracking systems, welding machines, flux recycling system, cooling system, and the temperature monitoring system near the heat-affected zone.

Success stories. Fully welded ball valves

Operating panel.
Following our philosophy of ergonomics and simplicity, we made a simple and easy to use interface on the robot's Teach Pendant, which made the system easy to use yet fully controllable. The advantages of this approach are evident, the operator can take the Teach Pendant to his work place next to the piece being welded and can reach every single parameter in the system, he has the information of the state of the whole robotic system including welding current and voltage and can control any parameter including welding speed and robot's position.

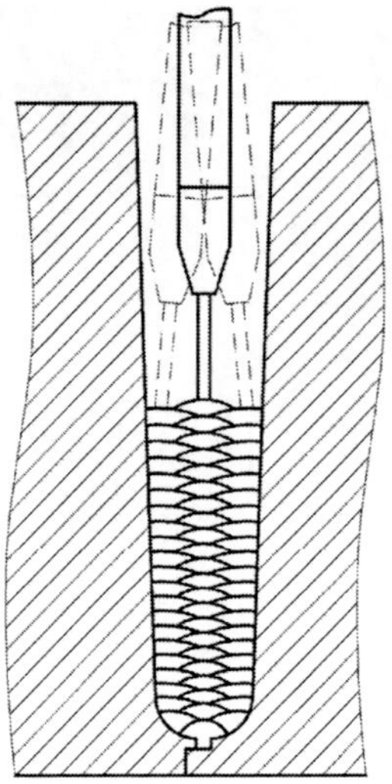

The welding process.
The welding process comprises gap filling layer by layer. Each layer has three passes. The robot has the ability to tilt the torch when approaching the walls of the gap. This allows to increase penetration and reduce the likelihood of defects. While filling the gap expands but for the robot that is not a problem because it can increase the cross sectional area of the welding pool. To ensure uniformity of heat input at the weld zone throughout the depth of the gap, robot increases the welding speed as the value of heat input increases. If for whatsoever reason two gaps are not equally filled, the robot can also change the cross sectional area of the welding pool to align the filling of two gaps. You can also carry out separate welds .

Advantages of this system.
1. The management of all elements and modules is performed from one controller and by means of one remote control.
2. The welding process is carried out fully automatically , the influence of the human factor is excluded.
3. The cooling system protects the internal components of the valve, and reduces the heat input at the weld region to improve the mechanical properties of the weld.

Success stories. Pipe branch tee

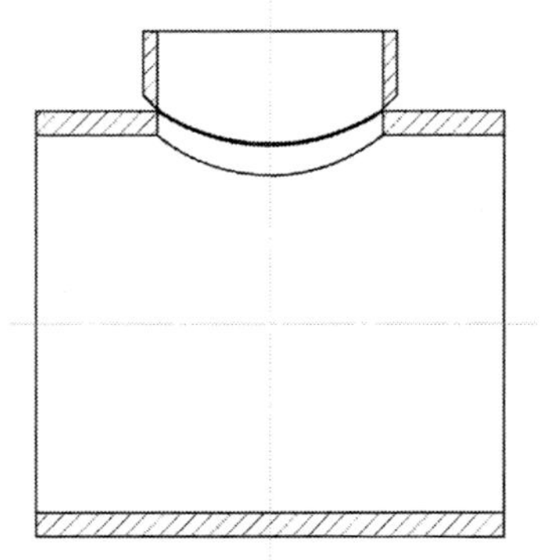

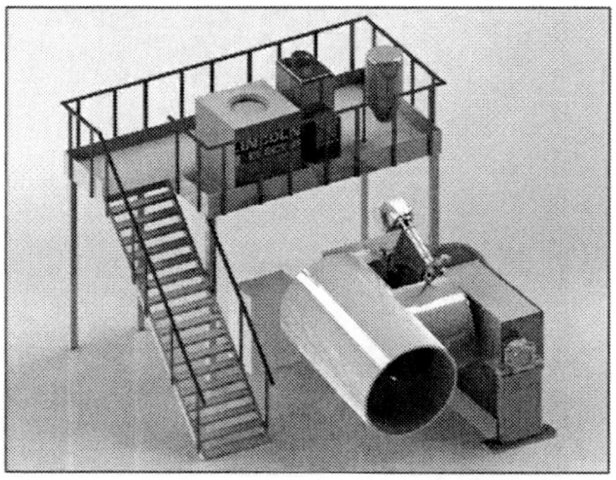

The use of robots for this kind of applications is not only convenient, but is a must. Saddle-shaped welds constantly keep changing the path and orientation for the welding torch, making it impossible for a standard 3 axis manipulator solving such tasks.
Our welding head together with the robotic system allow to reach and weld such gaps with variable profile.
At the moment we are working on the realization of this project.

Pre-study.

In preparation for this project, we have constructed an experimental platform where we carried out studies on welding such gaps and got positive results. To teach the robot, we have developed a special program that allows to define the trajectory of the passes and the gap's parameters within ten minutes' time. Based on this data, the system provides the optimum welding parameters which, if desired, can be adjusted. Next, the robot performs the welding in automatic mode.

Success stories. Pipe branch tee

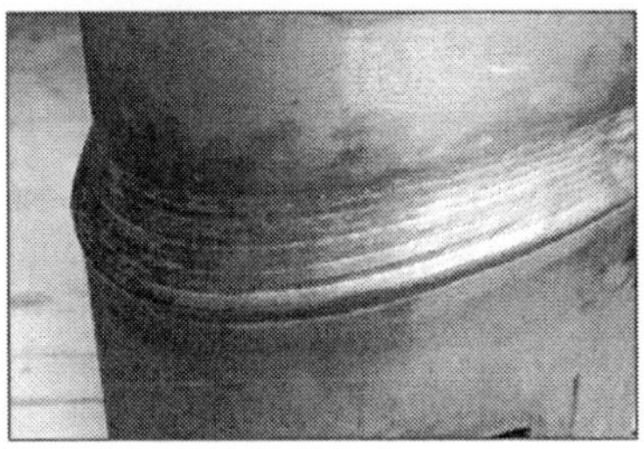

Know how.

ASG Robotics is a pioneer in these kind of welding. We have developed the unique technology able to weld these variable profile gaps, the equipment welds the metal in such way so the metal reaches the desired level uniformly at the last layer.

The advantages of welding such joints with this technology are evident: more joint strength, higher quality and lower costs at a higher productivity rate.

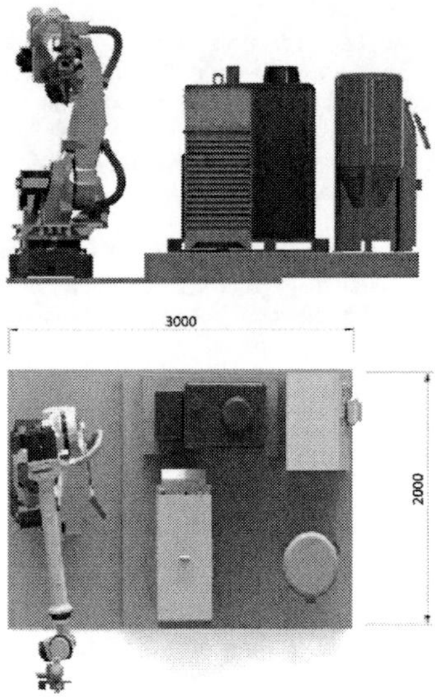

Based on acquired experience, we came to a conclusion that in order to make a widespread use of robotic submerged arc welding it is necessary to develop a universal mobile unit consisting of a platform on which the robot, welding machine, flux recirculation system, robot controller and wire are installed. To provide a larger working area we have chosen the robot R-2000iB/125L. The total weight of the platform is 5 tons. The platform can be moved by a forklift or a crane. For the accurate fixation of the platform relative to the work piece positioner or a linear guide for the longitudinal movement of the platform, high-precision mounting couplings are provided. This system is a complete unit ready for use. The system is scalable and can be equipped with a positioner, turntable, roller bearings or linear guide. We have developed a universal software, which only needs to be configured for a specific task. All these enables us to say with confidence that we can carry out welding of any circumferential or longitudinal welds of all types of joints.

Versatility of welding production systems by fixture-free GMA-welding and application of cooperating robots

U. Reisgen, A. Kampker and K. Willms, Aachen/D

Abstract

In order to meet the demands of the ever-increasing requirements of a competitive market, not only flexibility is required from today's production systems, they must, above all, be characterised by a high versatility. Versatility is to be understood as the capability to adapt system structure and technology very fast to changing boundary conditions, also in case of high diversity of product variants and/or low numbers of pieces per lot. For the quality-wise management of tolerances, the parts are, in automated welding production, mostly fixed in devices which themselves have often a highly complex design and are thus generating expenditure with regard to the manufacturing costs. Here, economic viability is given only for medium and high piece numbers, since the conversion to new products causes considerable changeover expenditure. In single unit and small lot production, therefore, manual welding is mainly used due to its high flexibility. Accordingly, in automated welding production, a high degree of versatility is only achievable if the final welding product is produced without using fixture devices from the primary individual parts. Therefore, the implemented approach to solution pursues the conversion of the welder-assistant principle of manual production to automated welding via the utilisation of cooperating robot systems. Within the scope of this presentation, the methods for the development of a versatile welding production system and the procedure regarding the implementation of the process sequence scenario "Welding without fixture devices" are illustrated.

Problem

Production systems are nowadays confronted with increased demands made not only to their flexibility but also to their versatility with regard to fast reaction to changes of economic factors, environmental conditions or customer requests. Here, the term versatility is to be understood as the capability to rapidly adapt organisation and production to changing boundary conditions beyond anticipated solutions. The versatility, however, should not be restricted to individual organisatorial and technological elements of production but should always aim for the versatile design of the overall system. This concerns, on the one hand, the value-added chain, and, on the other hand, particularly the internal business processes [1-2].

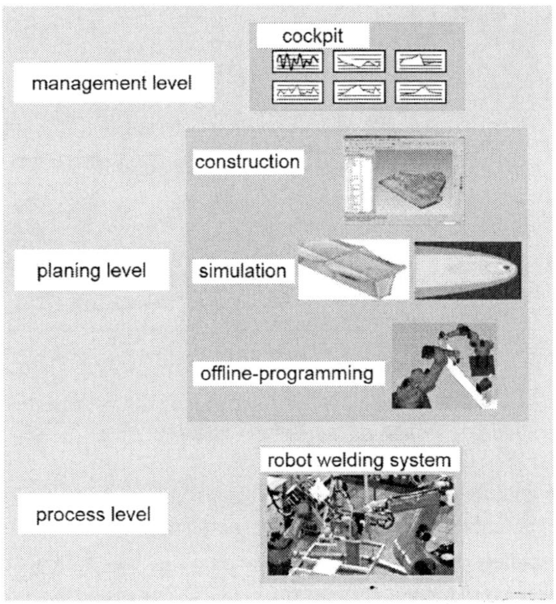

Figure 1: Solution concept in project ProAktiW

Therefore, the BMBF-research project ProAktiW aims for the development of a versatile production system on the level of individual enterprises via the integrated consideration of technology, organisation and information. In doing so, the design of the interaction of the individual elements on the management level and also on the planning and process level are in the fore (Figure 1).

Since versatility is accompanied by higher investment costs and technical effort, it is important to determine the necessary degree of versatility and the combination of solution elements in order to find the final versatility design for the overall system. On the process level, the research activities are concentrated on the automated welding production and on the search for suitable solutions for increasing the versatility.

Objectives of the Work

If gas-metal arc welding is used for manufacturing, mainly two production methods are applied. In the automated welding production, welding robots are used which are working on single or several work-pieces in one station. In order to control the tolerances, the parts are fixed in devices which are often technically complex and are thus causing high manufacturing costs. Although the use of a robot makes automated welding highly flexible with regard to the programming conversion to a new component, the fixture device is, as a rule, set-up specifically for the part and has a low degree of versatility. For this reason, economic production in automated welding is only possible in medium and large-scale manufacture. The conversion to a new product requires inevitably a new device which is connected with considerable changeover expenditure (Figure 2).

Manual welding is mainly applied in the fields of job shop and small-lot production. The fixture of the parts is carried out by a helper via fixing processes. Due to the flexibility and versatility of the operators, only little expenditure is required for changeover processes when a component must be changed. On the contrary, complex devices are a great obstacle for the versatility of automated welding manufacturing systems. A possible solution is the fixture-free welding. Under the term "fixture-free welding" in automated manufacturing, the set-up of products made of single parts via several cooperating robots without complex application-specific devices is understood. The compensation of the complex devices is carried out via application of the positioning ability of the robot system in combination with a suitable welding sequence. For the acquisition and compensation of geometrical variations in the joining zone, a light-section sensor is applied. For the implementation of the manufacturing scenario, the robots are working in intersecting work spaces.

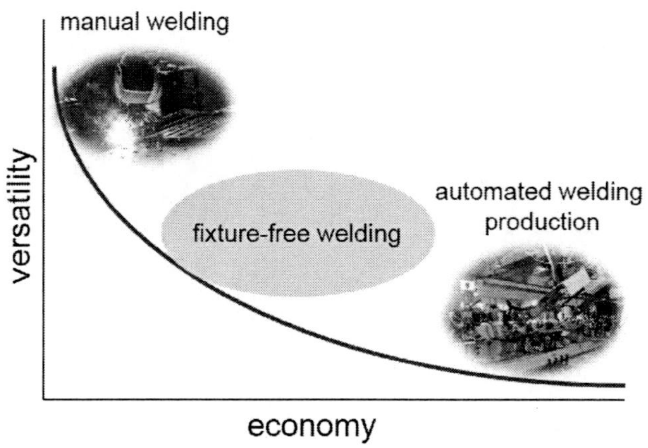

Figure 2: Classification of the solution approach of fixture-free welding

The solution approach shall be implemented in this project by linking the versatility of purely manual manufacturing with the economic viability of automated manufacturing. The aim is to manufacture components from the provided separate parts, if possible completely without using any clamping devices, which would make welding production also more attractive for low-volume production and small products. The vision about the solution of this problem is the transfer of the approach "Welder and Helper" in manual production to the application of cooperating robot systems.

Figure 3: Robot arrangement in the versatile welding manufacturing cell

In Figure 3, the robots of the welding manufacturing cell are depicted. The welding manufacturing system consists of a robot which serves as the "welder", of a robot which is equipped with corresponding gripping technique with a high ultimate load which shall assist as the "helper" and also of a turning unit as the platform for the set-up of the components. Already in the past, concepts for fixture-free welding with cooperating robots for the tacking of small attachment parts into readily welded components have been carried out [3]. We also know about solution approaches where robots are moving the parts under a firmly mounted welding torch where the parts are also fixed or at least pre-tacked in a device [4-6]. The complete set-up of the products consisting of individual parts has, so far, not yet been investigated.

The technical challenge in the implementation of the versatile welding manufacturing cell lies mainly in the transfer of the functions of a device to the cooperating welding robot system. This includes the positioning of the parts during tacking, the fixation of the parts partly also by design adaptations and the use of a suitable tacking and welding sequence for the compensation of the heat distortion.

Manufacturing Scenario for Fixture-Free Welding

For the development of manufacturing scenarios and methods of a versatile welding manufacturing cell, prototype parts made by industrial partners have been chosen. First tests have been made with a protection fence door which, for variety, consists of different fence elements of different sizes. The task includes the welding of the outer square frame and of the protective grid. During the positioning of the long frame parts, inaccuracies of the grip position may result in great variance in the joining area. The approach to solution may be either part-specific grips, the use of sensors or design changes at the part for fixture of the separate parts. Special sequence scenarios have been developed. In the following, a manufacturing scenario is presented where design adjustments at the parts which are to be joined are planned which would allow transfer of the device functions to the product. As a platform for the set-up of the fence elements, a turning unit is used. On this platform, the short frame profiles are positioned by means of angled stops in spacing direction.

For the compensation of the variations during positioning of the long frame profiles towards the short frame profiles, the use of additional gusset plates which are arranged in the corners of the frames is planned. The gusset plates serve, connected with the short frame profiles, as contact faces for the positioning of the long frame profiles and are thus reducing the positioning problem to one level. The complete manufacturing scenario for fixture-free welding using cooperating robot systems is depicted in Figure 4. The manufacturing scenario comprises all handling tasks and also tacking and welding processes.

In a first working step, the short frame profiles are positioned into the stops of the set-up platform and clamped with pneumatic toggle clamps. Then, the gusset plates are, successively, positioned into the recesses of the profiles with the taps, then tacked and welded. The turning unit is afterwards turned by 180° and the two long frame profiles are, one after the other, positioned on the gusset plates and tacked. In accordance with the welding sequence, now the corners are welded using different types of joints. As a last step, the grid is positioned, tacked and connected with the frame with short welded seams at the corresponding webs. Finally, the finished fence element is taken off the platform and another process can be set-up.

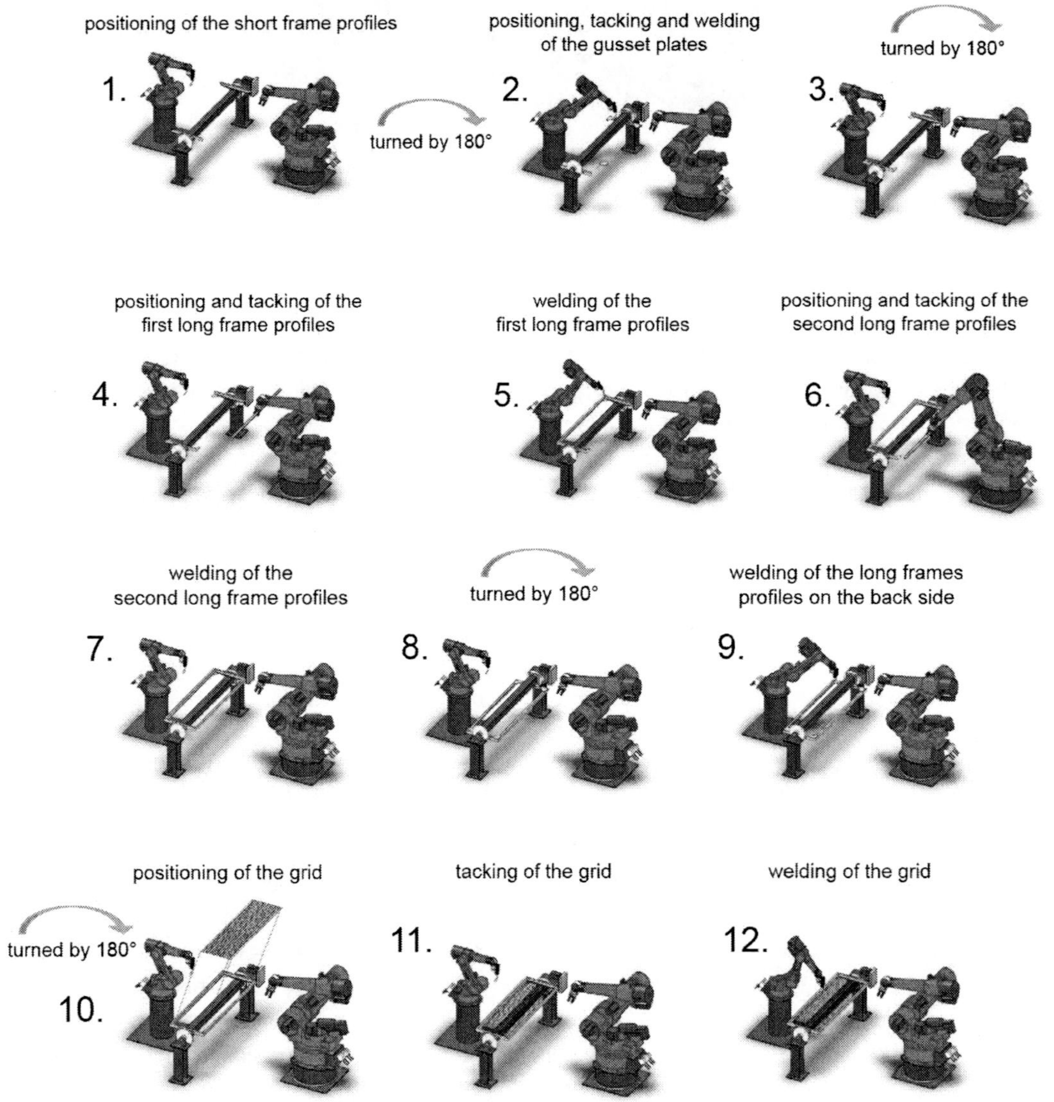

Figure 4: Manufacturing scenario for fixture-free welding

For ensuring the production sequences which are to be designed, the application of virtual planning is a promising approach. Here, simulation allows to carry out the evaluation of the technical and economical expenditure and the cycle times via the prepared manufacturing scenarios before the processes start [7]. Figure 5 depicts the visualisation level for the graphical simulation and the sequence plan for the motion sequences of the individual robots and also the required time.

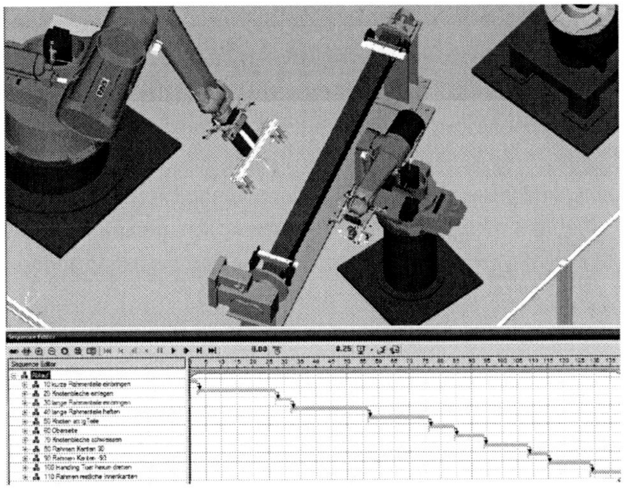

Figure 5: Offline programming and graphical simulation of the welding sequence scenario

Provision of System-Usable Welding Expert Knowledge

Another important point to increase flexibility in welding production is the provision of system-usable welding expert knowledge. Depending on the task and on the requirements made to the welded joints, this is mainly about the provision of welding process setting parameters on the planning and offline level and also on the process level. For generating the welding technology data it is, currently, common practice that in the event of part spectrum changes or variations a system operator with appropriate welding knowledge determines the required weld parameters for each task in time-consuming empirical weld tests.

In order to ensure versatility, a method has been developed which provides welding expert knowledge in a machine-readable form. The basis of this approach to solution is the use of surrogate models for the mathematical description of the connections between the setting parameters and the resulting weld qualities. Figure 6 depicts the procedure for the generation of the surrogate models. In a first working step, the definition of the welding task with possible, changes of the process boundary conditions, e.g. variation of the gap width, is carried out. The next step is the drawing-up of a corresponding weld test schedule. Here, the advantages of statistical design of experiments should be used since it allows to reduce the required welding work to an acceptable number but also allows to record, the influences of parameter combinations on the welding result. Subsequently, the welding results are subjected to evaluation via macro-sections.

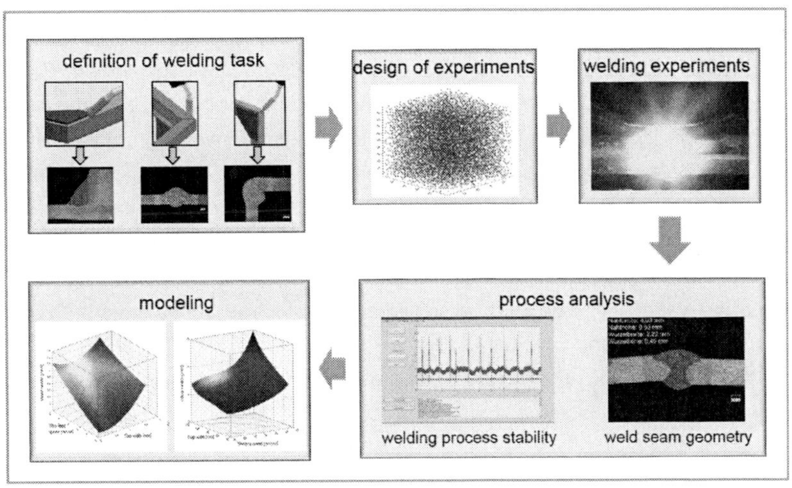

Figure 6: Procedure about the generation of surrogate models in welding

A further evaluation criterion is the process stability which provides information about the regularity of the weld seam formation and about frequency and distribution of weld spatters. For the evaluation of the process stability, the transient current/voltage signals are recorded during the welding tests. Further on, a tool has been developed in the ISF under MATLAB which allows to carry out automatically a stability evaluation for different welding process variations.

The generation of the models and also of the tools for model quality evaluation was carried out using the programming language R. Figure 6 depicts, by way of example, generated models which document the dependency of the weld width from the gap width in combination with the welding speed and also with the wire feed speed.

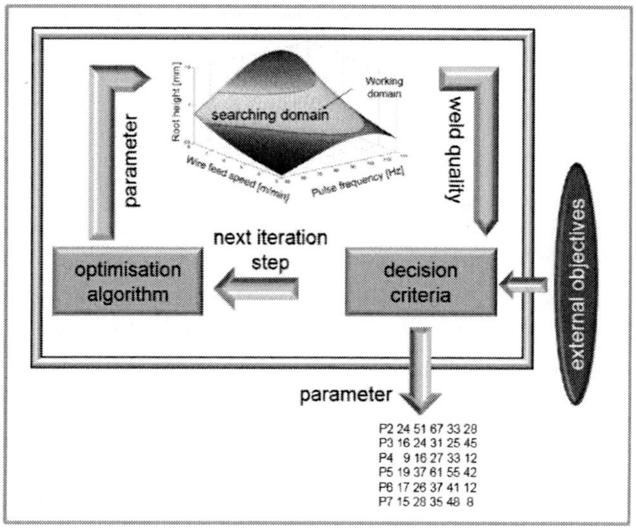

Figure 7: Optimisation methods for the generation of weld jobs [8]

The created models form the basis for the developed software tool for the generation of welding parameters, depending on the required weld seam quality (Figure 7). As objective, for example, the required weld seam geometry with a certain gap width is set. The optimisation of the welding parameters is carried out iteratively and uses a genetic algorithm. The evaluation of the results with regard to the objectives is carried out via decision algorithms. In the case of positive evaluation, the determined welding parameters are given out in the form of parameter lists which can then be used in a welding power source.

Summary and Perspectives

The research work which is carried out within the scope of the BMBF research project ProAktiW aims for the development of a versatile production system. Here, results of the works for the virtual planning and real production levels for increasing the versatility are presented, particularly with regard to welding production. Since in automated welding, complex devices are the greatest obstacle for versatility, a fixture-free welding manufacturing cell using cooperating robots has been developed and implemented. The functions of the device are, here, transferred to the robots in connection with a production scenario for the complete set-up of the product. The manufacturing scenario forms the basis for the virtual implementation and the assessment of the economic expenditure in a simulation system. For increasing the versatility, moreover, a software tool which is based on surrogate models has been developed and allows for the provision of system-usable welding expert knowledge. Depending on the task and on the requirements to the welded joint, it is possible to provide for welding process setting parameters on the planning and offline programming levels and also on the process level.

Funding Note

This research and development project is funded by the German Federal Ministry of Education and Research (BMBF) within the Framework Concept "Research for Tomorrow's Production" (fund number: 02PR2020) and managed by the Project Management Agency Forschungszentrum Karlsruhe Production and Manufacturing Technologies Division (PTKA-PFT).

Literature

[1] Westkämper, E.; Zahn, E. (Hrsg): Wandlungsfähige Produktionsunternehmen. Das Stuttgarter Unternehmensmodell. Springer. Berlin 2009.

[2] Wiendahl, H.-P.: Betriebsorganisation für Ingenieure. Carl Hanser. München 2005.

[3] Wihsbeck, M.; Wohlgenannt, M.: Vorrichtungsloses Schweißen verkürzt die Fertigungszeit drastisch. Maschinenmarkt, 2009

[4] Franke, D., Frick, K., Holm, H. und Moor, T.: Kooperierende Schweißroboter im Schiffbau – eine Laborstudie. UNIFORSCHUNG, Forschungsmagazin der UniBw Hamburg (10) 2000, S. 3-8.

[5] Widmann, W.: Vormarsch der Spar-Roboter - Kooperierende Roboter flexibilisieren die Fertigung. technik+EINKAUF 4, 2005, S. 50-51.

[6] Kuhlenkötter, B.: Kooperierende Roboter: Mehr Wertschöpfung in der Fertigung - Roboter arbeiten zu wenig. Industrie Anzeiger 23, 2008, S. 40.

[7] Pan, Z., Polden, J., Larkin, N., Van Duin, S., Norrish, J.: Recent Progress on Programming Methods for Industrial Robots. 41th International Symposium on Robotics ISR, München 2010

[8] Reisgen, U.; Purrio, M.; Buchholz, G.; Willms, K.; Lose, J.; Perge, J.; Schmitt, R.: Surrogate Modelling as an Enabler for Self Optimisation for Production Processes, CARV 2011: 4th International Conference on Changeable, Agile, Reconfigurable and Virtual Production; Montreal, Canada, 2 -5 Oct. 2011, pp. 642 - 647.

Use of EBW technology for interlinked production lines of the automotive industry

K.-R. Schulze, Neuberg/D, and J. Rugh, Enfield/US

It is widely known that any EB welding machine itself can operate fully automatically. But, it is also verified in many applications that it can be linked within a manufacturing cell or line to other processes needed in automated mass production either pre or post EB welding. And of course, loading and unloading the parts are done by automated handling systems. Besides quality demands, which are met by EB processes as it is well-known, floor-to-floor times short enough to satisfy the automotive industry's demands can also be realised. This is achieved by the use of EB machines having very short evacuation times (lock-chamber types) or needing no evacuation of a chamber at all (Nonvac types). The high flexibility of the electron beam is based on the very fast control of magnetic fields (focus, deflection et al.) and can be used to gain technical effects which are not easily achieved with other beam technologies. Moreover, magnetic fields cannot be polluted by vapour or spatter and beam output itself is stable - so EB process performance is easily ensured for long periods of time. Also, the operating costs of EB machines are lower than those of other beam technology installations. Finally, floor space requirements, clean factory demands and noise level limits of today's automotive industry are all fulfilled by automated EB manufacturing lines.

1. Electron beam welding – the underestimated technology

Unfortunately, it is widely believed today that electron beam welding (EBW) is a special process not suited for fully automated mass production. It will be shown here that this is not true.

EBW is characterised as having the lowest energy input per weld length and therefore produces minimum distortion of parts. This is true for thin as well as thick cross sections. So it is the right means to join finish machined individual parts to form a component which is ready for assembling. Moreover, different metal combinations can be fusion welded in order to produce a part with optimum performance characteristics. CNC-controlled EBW is generally done on parts precisely prepared, but tolerances or thermal deformation during welding or residual magnetic fields could influence the quality – especially the beam-to-joint positioning precision. However, this can be overcome by using the EB itself as a joint detection sensor – offline or online.

The attributes of EBW are generally known to engineers involved in the manufacture of large complex structures such as aircraft engines, power generating turbines and nuclear components. These items are manufactured in a single-item manner using stand-alone machines which mostly have large working chambers often having long evacuation times. Rather than using a general purpose machine, mass production EB welders, especially in the automotive industry, are designed to weld specific parts as efficiently as possible. Chambers are optimally sized and evacuation time is minimized. Fast and intelligent motion systems are also used in order to synchronise the EB cycle time (floor to floor) to the flow process of the superior production line.

The most common type of mass production EBW machine is the rotary indexing type (**Fig.1**). Two stations, one for loading/unloading and one for welding, respectively, operate alternately (in special cases four stations can be employed). Cycle time can be accelerated by using a load-lock type of chamber. Moreover, each station can carry one or more parts to be processed within one vacuum cycle – it depends on the size of the parts and the machine's general layout.

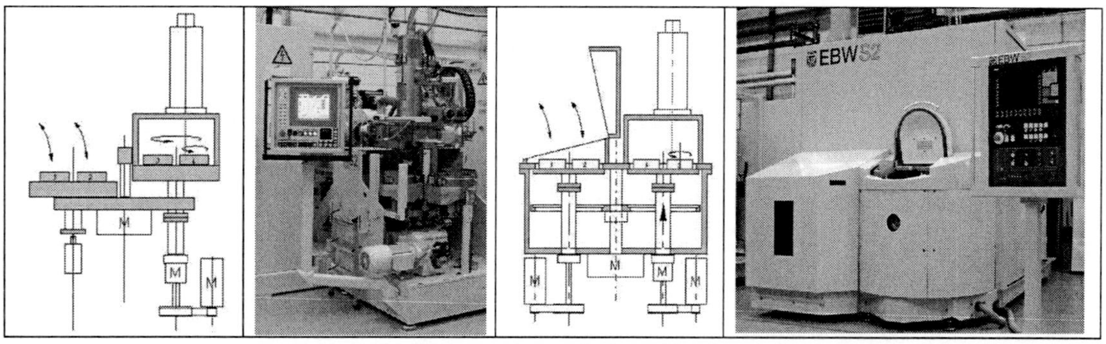

Fig. 1: Rotary indexing EB machines; **left:** the "open" P-type, **right:** the lock-chamber S-type [1]

By the way, these machines can be operated (loaded/unloaded) manually too, but they are best suited for automated interlinking. Of course, other machine designs are also available for this.

Low cycle times are not only guaranteed by fast evacuation (a few seconds) and chamber indexing, but also by an intelligent beam application. It is known among insiders that fast beam deflection allows splitting the energy input of one beam onto different impact points simultaneously (multi-pool welding) or into different energy distribution fields (multi-process welding). By using such means the EB welding process time can be shortened. Alternatively, the scope of EB processing can be extended to heating and/or smoothing in one single run. And in the case of two welds located wide away from each other on the same part, EB welders can be equipped by two generators operating at the same time.

Based on the excellent performance of an EB, very deep welding is possible in only one pass with butt joint preparation. In contrast to this, Laser beam (LB) welding - at comparable power - is restricted to much lower depths of penetration or demands a second process (hybrid) or multiple passes to fill the groove – not to mention the consumption of filler and gas. Nevertheless, LB welding is also used in the automotive industry – mainly in sheet joining for bodywork but also in gear manufacturing as long as cross sections to be joined are thin. But, in many applications EB welding shows a superior performance.

Often people argue that protection against X-rays would be a problem complicating the use of EB in daily work. As it is evident in many hundreds of installations worldwide – this is wrong. The design of any EB machine guarantees a human-safe operation without any additional effort. This also holds for use in mass production. By the way, LB installations without a vacuum chamber also need considerable protection enclosures, the manner of which depends on the wavelength and the power of the Laser light used.

An introduction to EB technologies and machines is given in a number of publications, e. g. [2; 3; 4]. These provide a good means to teach welding and production engineers as well as designers and managers, too.

2. Interlinking of beam welders in automotive mass production

As we can see in many examples, an EB machine operating in automated mass production is interlinked to a number of other units – either in line type or in cell type configurations. All these other units are also necessary in for LB - as long the task of the line (cell) is completely the same. In other words, it is a false assertion that the use of Lasers in automotive mass production would be much easier and less expensive than that of EB.

This EB to LB comparison also holds true with respect to the floor space consumed for the line installation. Whereas, the units upstream and/or downstream from the welding machine have to be all the same regardless of the type of beam used, the "footprint" of the welding machine (don't forget the chiller in case of LB) is comparable too. One issue sometimes raised by LB proponents is the need for part demagnetisation before EBW but not for LBW. This needs a floor space of half a square meter only or it is an integrated function of the preheating unit which may be necessary both before EBW and before LBW for metallurgical reasons of the work.

As mentioned already, the EB machine itself can operate fully automatically. The machine can also be loaded and unloaded automatically – often by robots. Transport within the interlinked line or cell is done by robots too, or by transfer chains (conveyers) or other means, **Fig. 2**. Joining a pair (or more) of individual parts requires identification systems to ensure that the correct combination occurs – from the first feeding up to joining.

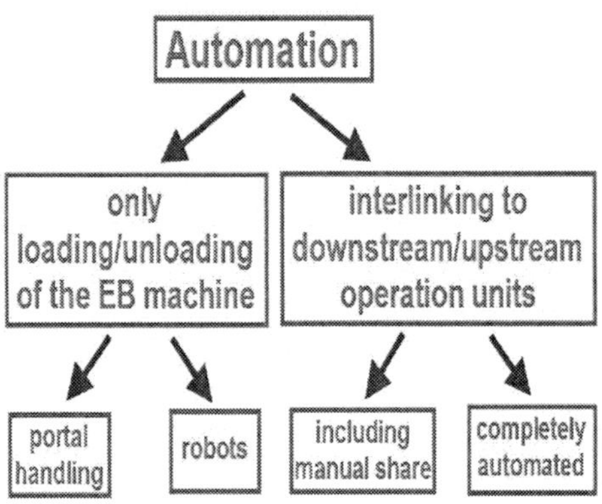

EB welders and automated production lines

- automated EB welder
- automated loading and unloading
- interlinking with stations before and after EB welding:
  - individualising and feeding of parts
  - cleaning and drying
  - assembling (pressing when required)
  - preheating (when required)
  - demagnetising
  - brushing
  - quality check (ultrasonic, optical, mechanical)
  - palletising
- all-inclusive safety features and equipment

Fig 2, left: Different stages of extension in automated mass production; **right:** constituent parts of a line

Keep in mind that parts coming from machining (turning, milling et al.) must be cleaned and dried in order to avoid weld defects. Tunnel type cleaning machines are normally used in linked lines, instead of batch machines.

Assembling of parts is possible in different manners; it depends on the nature (design) of the product. In some cases single parts are put together in the EB machine directly and clamped by the tooling. In most cases however the part design contains a press fit (which ensures a zero gap which is preferable for EBW). Then fitting is done outside of the EB machine under controlled force-displacement characteristics.

As mentioned before, residual magnetic fields within parts will cause welding defects if the fields are stronger than 5 Gauss. To go below this level, all steel parts are demagnetised before inserting them into the EB machine. In the simplest manner this is done by moving the part through a strong magnetic coil with alternating field whose strength decreases with distance. If the part needs preheating before welding because of metallurgical reasons, a very effective demagnetisation is obtained as a by-product of the uniform magnetic heating process. By the way, this very fast acting low-frequency technology was introduced many years ago by PTR to EB welding of gear parts in order to replace the long-standing oven process which had constrained fully automatic mass production lines in the automotive industry.

After welding some additional operations may be necessary: brushing-off spatter, quality inspection and finally handing over the parts to the next station in the line. Regarding spatter, EBW can be optimized for many automotive parts so that it produces minimal or no spattering; so most installed brushing stations are often out of action.

Quality inspection is normally integrated within the automated EB cell. A 100% ultrasonic check is very common in the automotive industry; other checks are video survey or laser measurement of the top bead.

Let us have a look onto some examples of interlinked mass production using EB technologies in the automotive industry. This mostly concerns components of the engine and/or the drive chain; but it should be emphasised that there are many more such applications which go beyond the scope of this paper.

3. Gear welding by electron beam

There are many different part families for manufacturing car or truck gear drives – small or big, manually or automatically shifted, front or rear, longitudinally or transverse, etc. Some of them consist of two individual parts only; others are joined from more. **Fig 3** demonstrates a number of automotive parts which are EB welded at different places around the world – all in interlinked automated production lines.

Fig. 3: EB welded automotive parts for gears and engines

Fig. 4 shows the layout of an EB welding cell for gear parts – an output of 1 part every 15 seconds is proven. In this case (as is often the case) the same machine or line has to produce different types (or sizes) of a given part family. Because of this, clamping tools, grippers and other means must be adjustable – preferably with minimum or no effort. It is not a great idea to feed such a line by a chaotic flow of parts, but it would be not impossible.

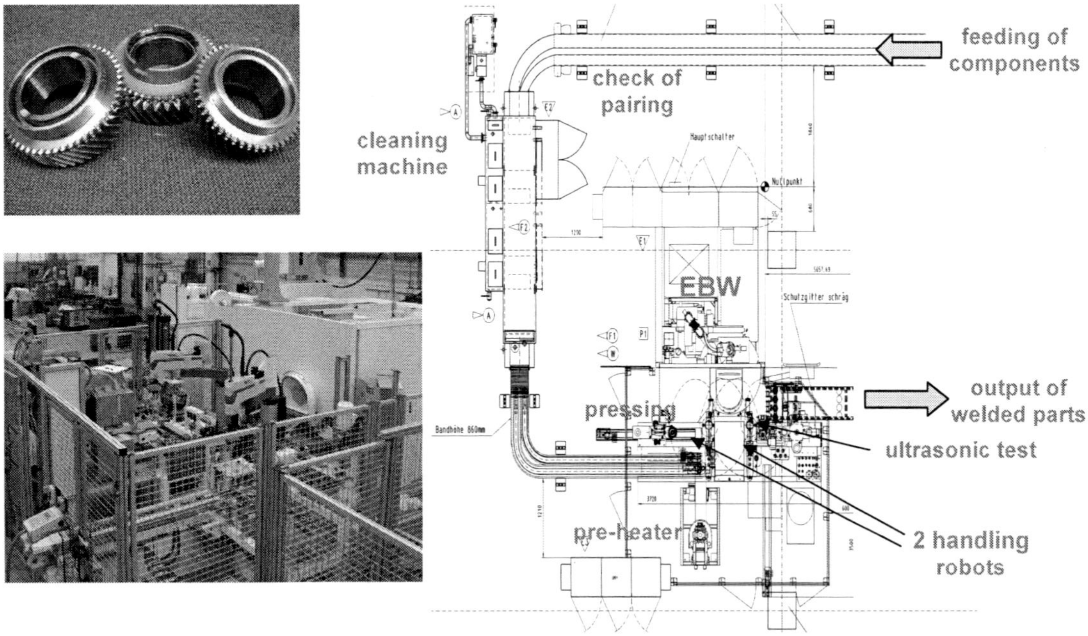

Fig. 4: Interlinked EB cell for welding of gear-wheel and synchronising ring – here for gears 3, 4, and 6

The respective layout of any cell or line takes into account both the technical demands of the approach to be realised and the needs of the user and the plant organisation. So it will look quite different for different applications – apart from replicating any cell in order to enlarge the user's output.

There are many other installations all around the world for EB welding automotive parts in large quantities. And some of them are not equipped with robots (as given in the example above), but are using portal systems for handling operations. This depends on some preferences of the user and is realised by the EB-machine supplier – sometimes in cooperation with the customer's staff [5].

Because of the complexity of such cells, it is not a good idea to make its footprint as small as geometrically possible. The layout should respect the access and freedom for service activities too.

4. Local surface hardening by electron beam

To show a second example we refer to another very successful EB approach: surface hardening. Besides otheres, this is done on AVS cam shafts after final grinding of the contact surfaces, **Fig. 5**. By fast deflection of the beam within a certain field and simultaneous turning of the shaft, heat is locally applied to the oval track so that a material transformation takes place up to a certain depth, first into austenite and thereafter into martensite by self-quenching. There is no need to externally cool the part to harden it. Unfortunately, it comes out from the EB process at too high a temperature to be allowed for the following operation: oil finishing.

Therefore, the EB hardened part must be cooled down and water was found to be the most effective means. Clearly, after cooling the part with water it becomes wet, this is also forbidden for the subsequent oil bath. So, blowing it dry by air – all around, outside and inside – is unavoidable.

As if not complicated enough, each part has to be identified – both incoming and outgoing – by reading a data matrix code (DMC), and the applied EB process parameters have to be written by Laser marking onto the individual part. All these operations are interlinked within a cell (**Fig. 5**) which is connected to the factory's conveyer by a satellite band.

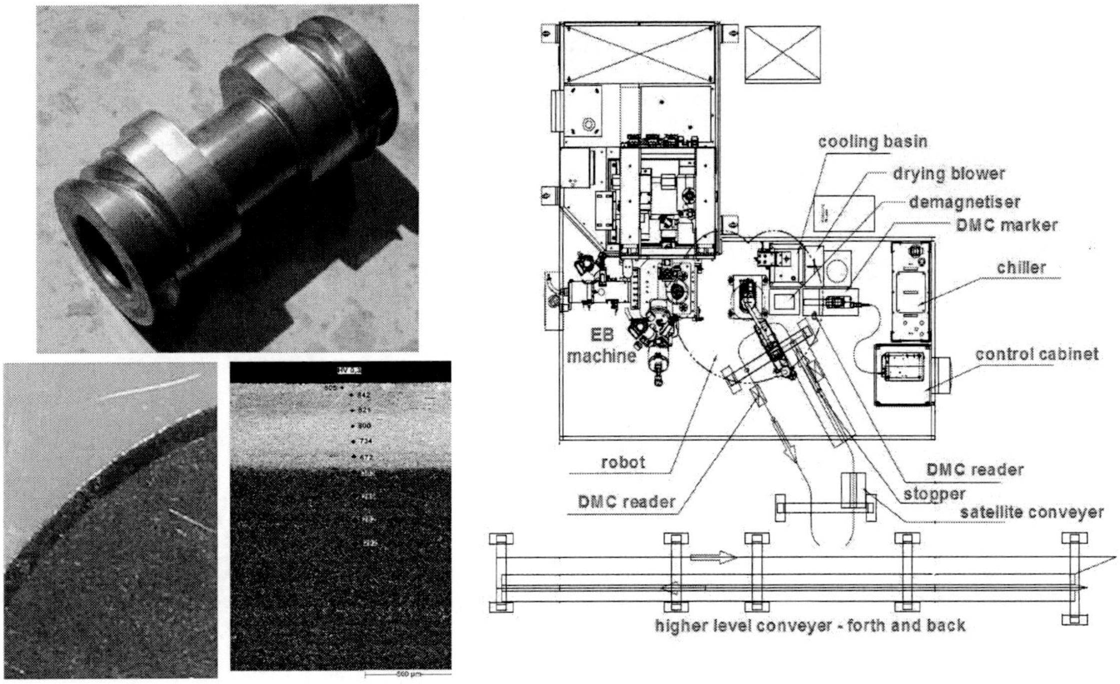

Fig. 5: Interlinked EB cell for surface hardening of cam shafts – 4 tracks per part; coping with different shaft types

The productivity of this cell amounts to 14 seconds per part. This could be achieved by using a special design of a small-sized load-lock indexing machine.

By the way, there are concepts to combine welding and surface hardening (of turbocharger rotors) in one operation using the same machine and automatic part handling [6].

5. Nonvacuum EB welding of car body parts

In contrast to the examples given before which apply the electron beam in vacuum, the Nonvacuum variant (NVEBW) does not need a vacuum chamber for part processing (of course, the beam itself is generated in vacuum). So, especially large parts can be welded very quickly as long as the cross section to be joined is limited. This is the case with the so called cross-beam which is placed between the A-columns of a car and carries a number of modules. In addition, a family of different designs has to be manufactured depending on the car model. And all those variants have to be welded in the same cell (**Fig. 6**).

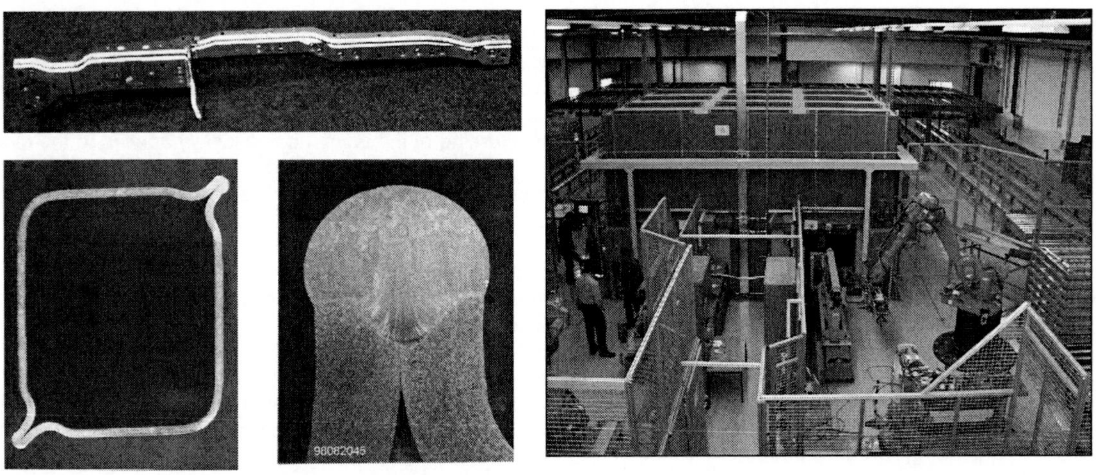

Fig. 6: NVEB welded module carrier and double installation (2 housings back to back) of fully automated cells

In general a module carrier consists of two deep-drawn half-shells of AlMg3 sheets, 1.5 – 2.5 mm thick. The weld is done as a flange seam along the 3D contour 1.5 m in length – each on both sides. There is no need of additional material (filler or gas), and the EB process runs without any difficulty in spite of the aluminium material. Moreover, remarkably large tolerances within the sheet assembly are bridged by the wide Nonvac beam. One weld is accomplished within 8 seconds since the NVEBW process is very fast (too fast from some people's point of view). The individual parts are fixed in a clamping unit which is adapted exactly to each model.

Whereas, the EB process inside the protective enclosure (against X-rays) runs automatically – CNC moving the gun column as well as the clamping unit – assembling of the parts and loading/unloading the clamping unit is done by several robots outside the enclosure.

One (1) complete cell produces a welded part every 31 seconds. Only one manual operation is necessary – placing of the individual parts onto feeder slides from which a robot takes them for assembling. Also, a quality check after welding is done automatically: The unloading robot moves each weld along a camera to evaluate it and stores the part thereafter in a transport box.

6. Automation for mass production inside the vacuum chamber

In contrast to, and as addition to the given examples of interlinked automation of an EB machine within a line or cell, a very special automation inside the working chamber was realised in order to weld in unattended manner up to 230 parts loaded batch wise, **Fig 7**.

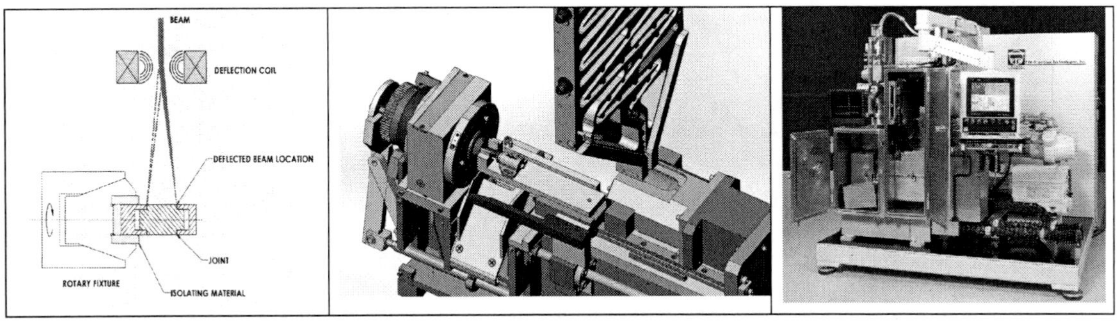

Fig. 7: Inside automation of an EB welder for cylindrical parts; **left:** scheme of the welding process, **centre:** magazine with parts (purple-coloured) and collet with loading tool (index wheel and slide); **right:** EB machine opened

This system was specifically designed to weld cylindrical parts that vary in size from 10 mm to 30 mm in diameter and from 30 mm to 90 mm in length. These cylinders have two radial welds which can be separated by up to 60 mm along the cylinder's length. The movement between these welds is accomplished solely by beam deflection (**Fig. 7, left**) and a positioning calibration routine verifies the accuracy of this deflection to within 0.025 mm using a specially designed faraday cup. After the system checks out OK the part loading sequence begins.

One magazine at a time is loaded into the upper portion of the right compartment of the vacuum chamber shown in **Fig 7, right**. This two compartment vacuum chamber protects the magazines and part handling mechanisms from the weld spatter, vapour and heat generated in welding side of the chamber. This is the reason for the unusual shape of this machine.

The parts in the magazine roll down the serpentine slot to the feeder wheel (**Fig. 7, centre**). The feeder wheel indexes 180 degrees to deposit one part into the part insertion yoke. The part insertion slide then transfers the part into the welding side of the chamber and inserts it into the collet. The part is precisely positioned within 0.025 mm into the collet by using opposing springs to assure the part is positioned against a locating face on the part. The insertion slide is then retracted from the welding chamber. Once the part is welded, the collet opens, the part is ejected into a bin inside the chamber, and the cycle starts again. The system can then cycle through the entire magazine of parts automatically without any operator intervention.

The production demands on this system require switching between part numbers almost every day so this system was designed to allow quick change over between the 20 different part numbers. To ease the changeover, the entire tooling assembly slides out of the vacuum chamber on rails then is slid back into position after the collets and corresponding loading devices have been changed (**Fig. 7, right**). This maintains the alignment of the tooling after a changeover.

Welding for 2 - 4 hours using up to 7 kW of power without venting the vacuum chamber can cause a significant amount of heat to build-up. Allowing this heat to accumulate in the magazine of parts to be welded or in the tooling could easily cause the welding process to drift out of tolerance over the course of 200 parts. To allow the process to run at steady state, active cooling is employed in the rotary fixture. This process has proven to be stable and able to meet the 0.025 mm dimensional tolerance requirements even though some parts are over 500° C when they are ejected from the rotary fixture.

7. Economy of EB welding – an essential aspect in automotive mass production

Beam welding in general – EB as well as LB – opens so many possibilities for saving costs, even in the design phase of components. They can combine finish machined individual parts, even those made of different metals, in order to develop a very efficient process or to gain special properties of the products. This is commonplace.

Especially in mass production, costs are decisive for the choice of this or that technique – besides quality and cycle time, of course. In the case of an EB welder, all aggregates needed for carrying the technology through are included within the machine itself; so investment costs can be calculated clearly. In the case of LB welding, often only the price of the Laser itself is considered in the cost comparison, which is incorrect. Because, to install a turn-key ready LB cell the safety enclosure, the beam guidance, the tooling and more have to be taken into account too – all together these can amount to three times the price of the laser alone. No question, both approaches are able to be interlinked in production lines and the layouts can be comparable.

Let us have a look to the running costs of an EB machine. The beam generation system contains one real consumable: the cathode. This little tungsten filament has a service life of umpteen hours of beam-on time – not clock hours (remember: if the EB is not in use – even for seconds – the generation is deactivated completely). Exchange of a worn cathode needs a production stop of 5 minutes in total, not more. And the price of a cathode amounts to less than 20 €.

What about the beam quality over time, especially after cathode exchange? Today's technique with fast deflected beams allows measurement and evaluation of the beam profile (energy density distribution across different levels) and adjusts it automatically – by changing magnetic fields only – to the reference of that beam which has been previously qualified. This procedure takes one or two minutes only. It has been used for some years within general purpose chamber machines performing welding jobs on valuable components, but now we have the opportunity to include it in index-table machines, **Fig. 8**.

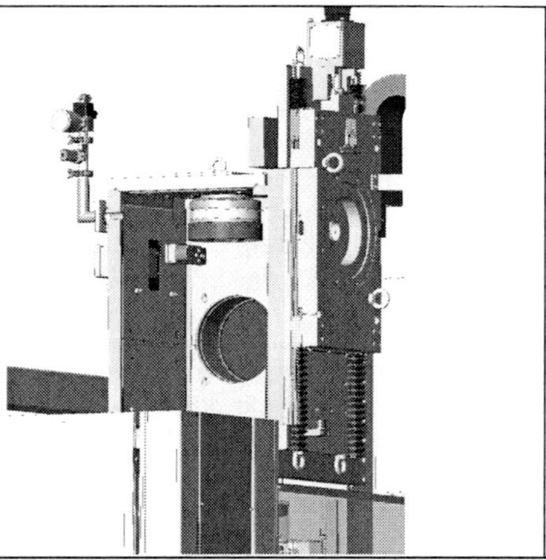

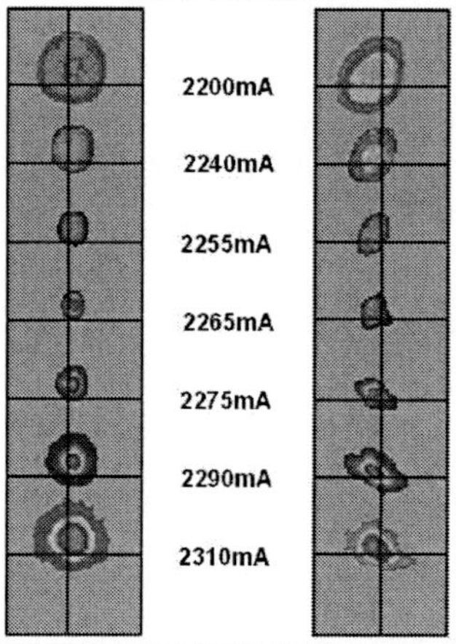

Fig. 8, left: Installation of a measuring system EBO-Set (pink coloured) inside an index-table EB-machine (horizontal EB generator will be placed onto the green coloured slide); **right:** Density distribution patterns of a well adjusted and an unusable beam, respectively, together with focus currents

It should be emphasised here that all means for forming and guiding the beam from the source until the workpiece are not subject to wear, because of its (magnetic or electric) field nature. This means that besides the cathode, no other wear parts have to be exchanged on an EBW machine, keeping operating costs low.

Well, there are some additional consumables: a very low amount of pressured air to activate evacuation valves, a few litres of oil for refilling a vacuum pump (twice a year), some seal rings to be replaced (once a year) and of course electrical energy. The latter one mainly determines the running costs of an EB machine.

We have measured the electrical energy consumption of a real EB machine (open index-table type, comp. Fig 1, for gear manufacturing): Wall-socket current draw as well as consumption of all components inside the machine. The result is illustrated in **Fig. 9**. The numbers obtained depend on the real welding cycle parameters, of course. The 43% relationship between wall-outlet consumption and beam power is based upon 6 kW applied onto the workpiece in this particular system. The relationship becomes better still as beam power increases – at 10 kW a number of 69% results.

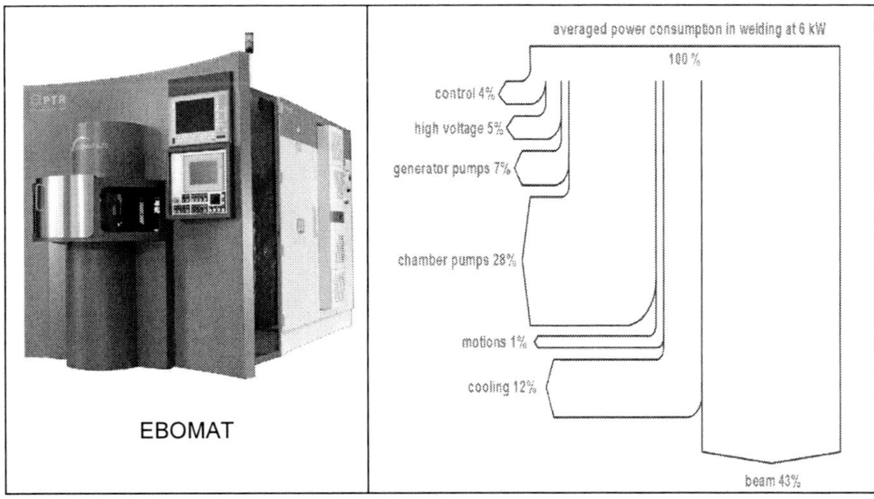

Working cycle regarded:

- Loading
- Evacuation
- Indexing
- Welding 15 s at 6 kW
- Venting
- Indexing
- Unloading

Total cycle time = 26 s

Fig 9: Chart of time-averaged power streams within a defined working cycle of an index-table machine

A qualitative comparison between EB and LB energy consumption is given within different former publications [7]. The EBW chart there is close to the quantitative result obtained by the measurements above; but unfortunately no quantitative data are given for LBW (they will be definitely different for the various types of Lasers). It has to be emphasised that the energy consumption of the EBW machine includes all components needed – from evacuation to beam generation and part movement up to system cooling. This should be borne in mind when LBW systems are compared.

8. Conclusion

It is quite evident from many applications that EB technologies are well suited for mass production within automated interlinked lines – not just in the automotive industry, but elsewhere too. Nobody would assume that it is just "freakiness" that big companies – in Europe as well as in Asia – have been operating umpteen EB machines (not just all together, but in each enterprise) for many years, most of them interlinked. And, the ramp up continues.

People who only have lasers in mind for this type of automated welding should be open minded and also look at alternative beam processes in future. The benefits of EB processing can be significant if the efficiency of the energy transformation from the beam to heat and the outstanding performance of the fast electron beam manipulation are taken into account. Electron beam welding is able to perform sound welds without inside defects and outside spatters – of course, some materials and/or designs demand skill in developing the right procedure and parameters. And last but not least, the low running costs of an EB welding system are very convincing.

By the way, especially for gear manufacturing, people are now working to apply laser beams inside a (soft) vacuum chamber in order to better the weld quality to a level proven day by day by using an electron beam...

9. Literature

[1] K.-R. Schulze: Hochproduktive Massenfertigung mit Elektronenstrahl-Schweißmaschinen vom S-Typ. Vortragsband 6. Int. Konferenz Strahltechnik, Halle/Saale, 2004, S. 56-62

[2] H. Schultz: Elektronenstrahlschweißen. Fachbuchreihe Schweißtechnik, DVS-Media Verlag, Düsseldorf, 2000

[3a] K.-R. Schulze: Elektronenstrahltechnologien. Wissen kompakt Band 1, DVS-Media Verlag, Düsseldorf, 2011

[3b] K.-R. Schulze: Electron Beam Technologies. Wissen kompakt Band 1e, DVS-Media Verlag, Düsseldorf, 2012

[4] AWS Welding Handbook, Vol. 3, Part 2, Chapter 13: Electron beam welding. Miami/FL, 2007

[5] U. Müller and K.-R. Schulze: Unterschiedliche Fertigungsstrategien in der PKW-Getriebefertigung mittels Elektronenstrahl-Schweißen. Vortragsband 7. Int. Konferenz Strahltechnik, Halle/Saale, 2007, S. 128-132

[6] K.-R. Schulze: Multitalent Elektronenstrahl – Fügen und Härten am gleichen Werkstück in ein und derselben Maschine. DVS-Berichte Band 267, Düsseldorf, 2010, S. 68-72

[7] DVS-Merkblatt 3209: Wirkungsgrade beim Elektronenstrahl- und Laserstrahlschweißen. Düsseldorf, 2007

CMT-Twin – A quantum leap to improve welding automation

H. Hackl and S. Egerland, Wels/AT

Introduction

/ Dramatic Shortage in Qualified Welding Personnel

/ Negative Effects on Weld Quality, Productivity and Industrial Competitiveness

/ Different Approaches to overcome the Situation:

- *Improving the Image of Welding*
- *Improving Welding Education (e.g. Virtual Welding, Online-Learning)*
- *Improving Welding Career Opportunities*
- *Improving production means*
- *Rising the degree of mechanisation and ...*

/ Automation as a considerable means to battle *both* welder shortage and weld quality decrease

Introduction

/ Classification for different degrees of mechanisation in welding[1]

- **Partly Mechanised Welding** - e.g. mechanised wire feeding
- **Mechanised Welding** - mechanically or electronically controlled welding parameters
- **Automatic Welding** - welding operations pre-set and performed automatically during welding

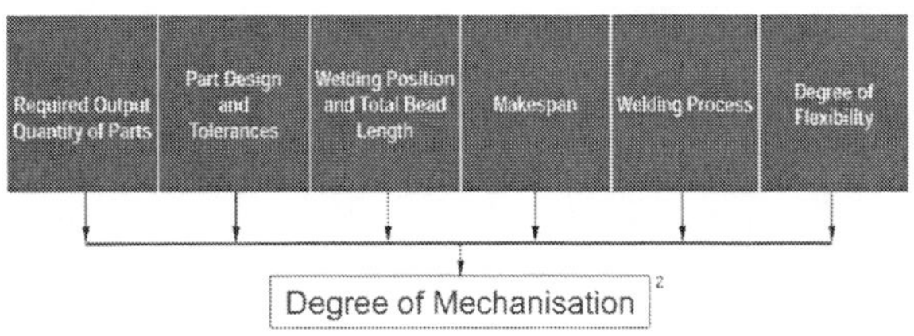

[1] British Standard BS 499-1:2009 Welding Terms and Symbols

[2] K.-P. Schmidt, "Vollmechanisches MIG/MAG-Schweißen – Geräte und Auswahl"

Advanced Welding Automation – Perspectives 2015

- Higher degree of mechanisation due to lack of qualified welding personnel
- Higher quality of automation solutions corresponding to generally improved technological advancements (electronics)
- Requirement for turn-key solutions from only one key-supplier
- 'Steel' as remaining *the* most important parent material
- Increasing number of composite application (transportation sector)

Advanced Welding Automation – Perspectives 2015

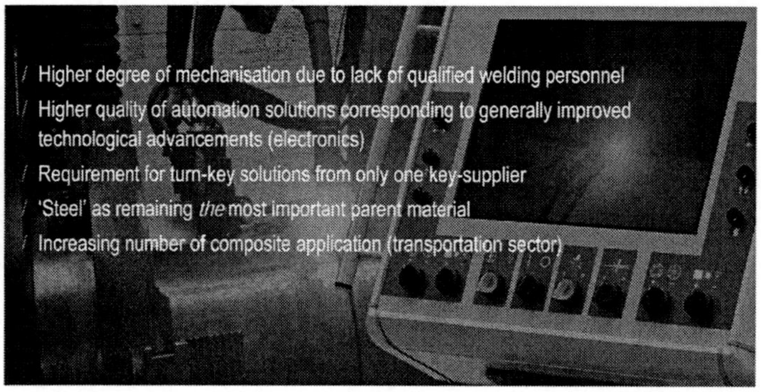

Advanced Welding Automation – Consequences 2015

/ Approx. 10% process automation growth in average worldwide
/ Decreasing product life cycles
/ Increasing number of articulated robots due to dropping capital cost, greater flexibility and benefit
/ Greater impact of quality control and monitoring (sensors)

Advanced Welding Automation – Requirements 2015

/ Higher Performance in both hard- and software
/ Shorter time-to-market cycles, i.e. stronger offline process simulation impact
/ Improved Quality Management (QM) and Data Logging

Advanced Welding Automation – Innovative Answers to the Future: *Compact Cladding Cell*

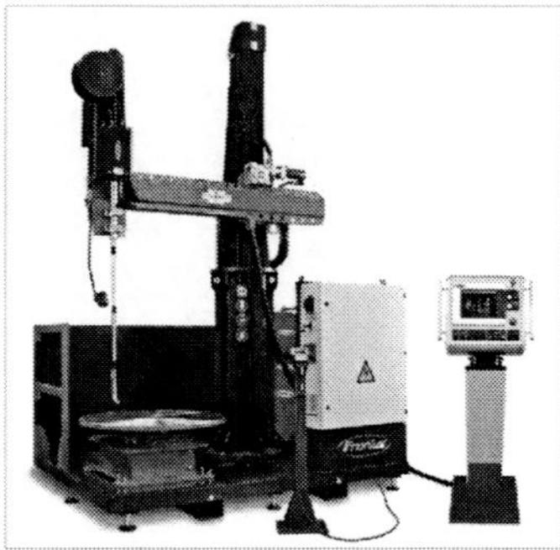

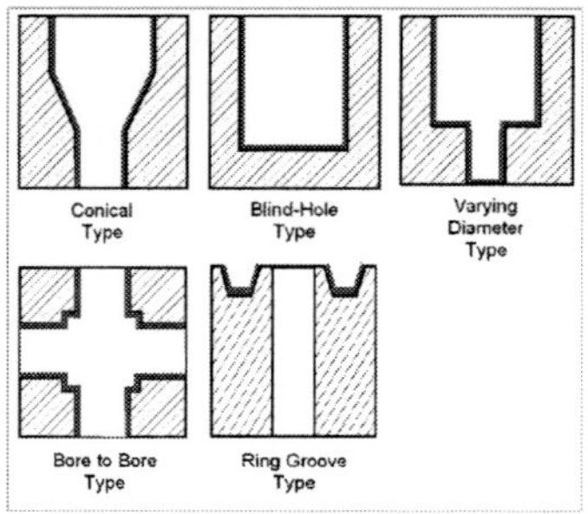

Advanced Welding Automation – Innovative Answers to the Future: *ArcView* Optical Sensor Solution

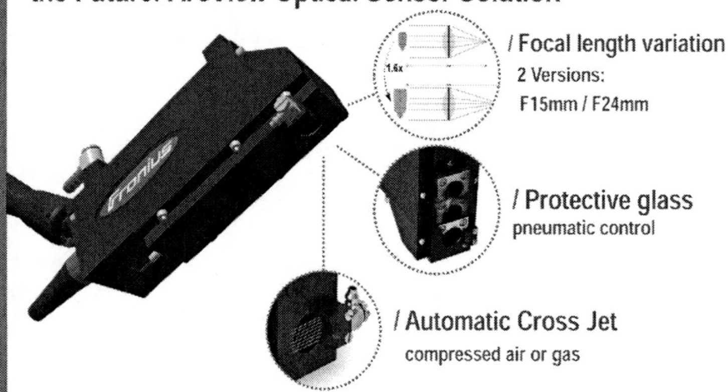

Advanced Welding Automation – Innovative Answers to the Future: *ArcView* Summary

/ Accuracy/Repeatability
/ Adaptability to Groove Change
/ Spatter Resistance
/ Thermal Resistance
/ Arc Irradiation Resistance
/ Fume Resistance
/ Shape/Size
/ Durability/Lifespan
/ Proper Maintainability

The Welding Process – Industrial Requirements

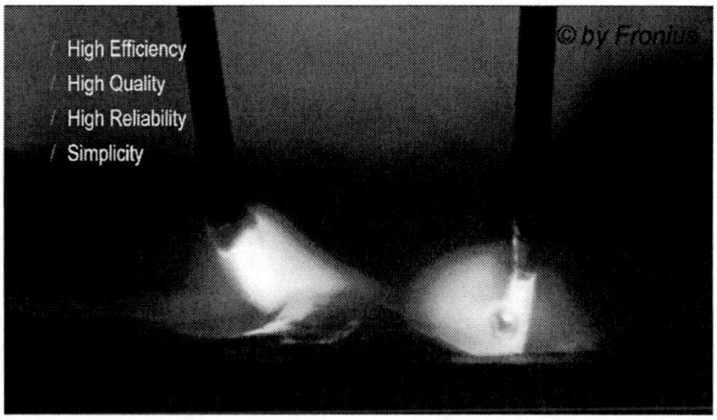

The Welding Process - Future Perspectives

/ Increasing Importance of Controlled Welding Processes (CMT)

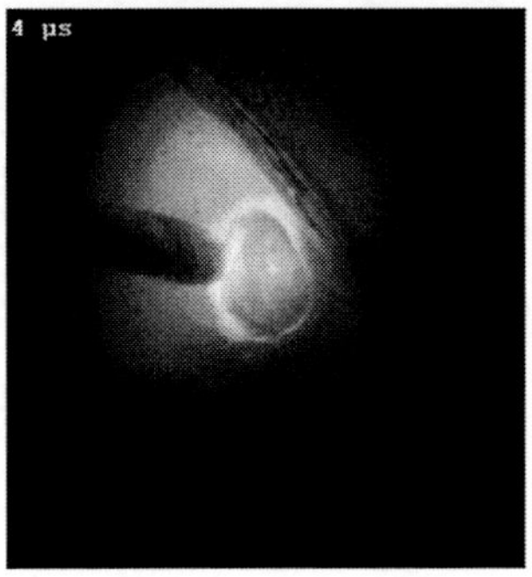

/ CMT Single Wire Welding

/ High Stability

/ High Result Reproducibility

/ High Weld Performance

/ Original CMT Welding Characteristic; i.e. high Current Peak shortly prior to Short Circuit Release

The Welding Process – Rising Efficiency in Automation Conventional GMA-Tandem Welding

CMT-Twin –Application Range

Variant (*Commercial Denotation*)	1. Wire Electrode (leading)	2. Wire Electrode (trailing)	Specific Requirement
CMT-Twin '*Speed*'	GMAW-P	CMT	Travel Speed
CMT-Twin '*Heavy Duty*'	GMAW-P	CMT	High Wall Thickness
CMT-Twin '*Root*'	CMT	CMT	Root Pass Welding
CMT-Twin '*CO_2*'	CMT	CMT	100% CO_2 Shielding
CMT-Twin '*Cladding*'	CMT	CMT	Surfacing

The Welding Process – Example: CMT-Twin *Heavy Duty*

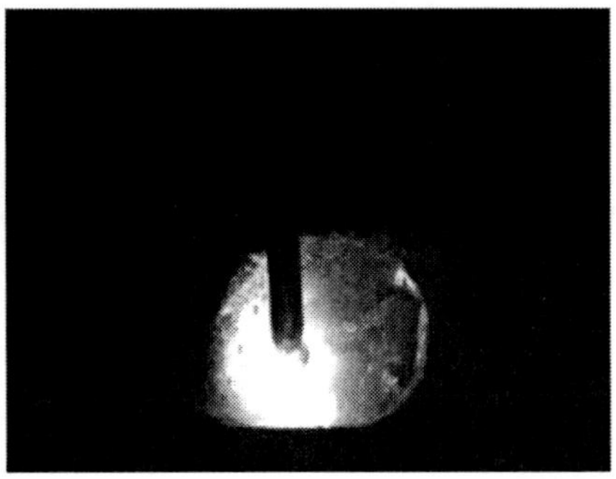

- GMAW-P + CMT process combination
- Remarkably reduced arc interference vs. conventional GMA-Tandem Welding
- Safe penetration + High Weld Travel Speed

- **Highest Quality Levels:**
  - Fusion Profile
  - Bead Shape

The Welding Process – Example: CMT-Twin *Heavy Duty* Arc Ignition

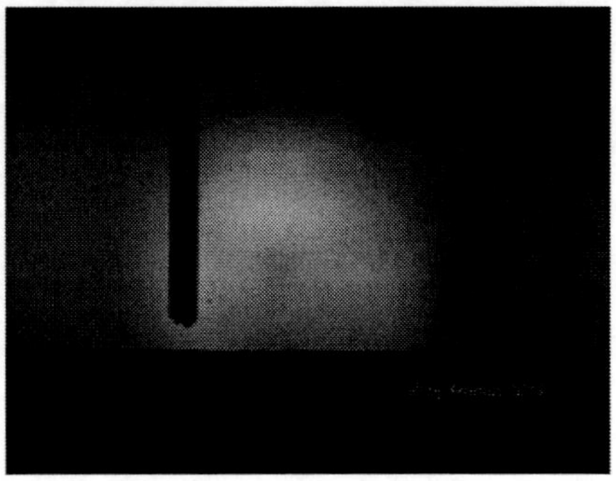

- Critical Arc Ignition Period in Conventional Tandem Wire GMAW
- Precise CMT-Twin Arc Ignition
- Beneficial quality improvement

The Welding Process – Example: CMT-Twin *Heavy Duty* Welding Application

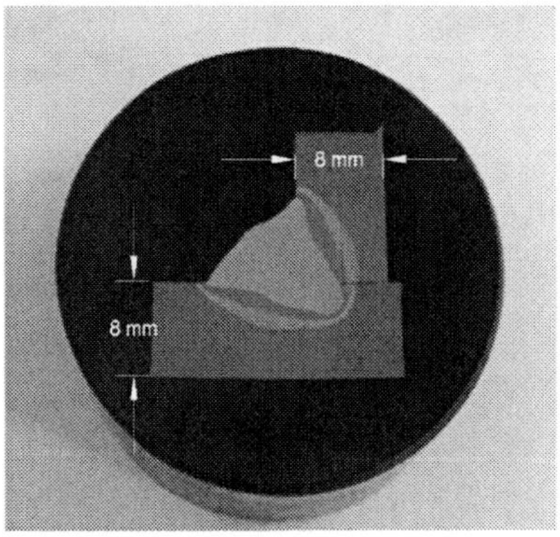

- **Parent Material**: Wear Resistant 450 HBW Steel
- **Thickness**: 8 mm
- **Filler Material**: EN 14343-A G3Si1 ø 1.2 mm
- **Welding Position**: PB
- **Weld Travel Speed**: 100 cm min^{-1}
- **Requirement**: Replacing ø 1.4 mm FCAW

CMT-Twin – Quantum Leap in Rising Welding Efficiency

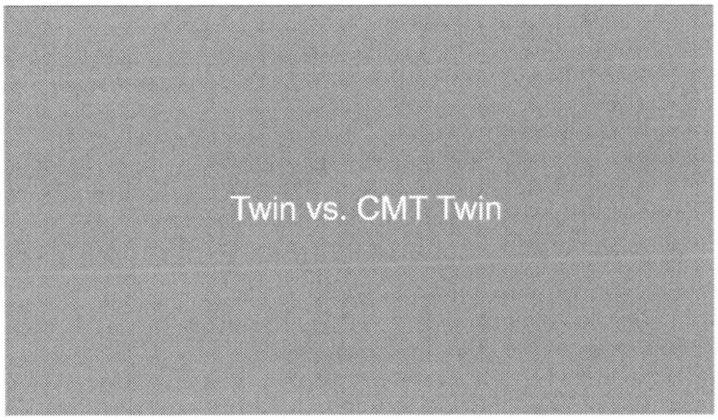

CMT-Twin Welding in Automation – Application Shipbuilding

- CMT-Twin *Heavy Duty* Travel Speed: 140 cm min^{-1}
- Excellent Root- and Cap Pass Shape
- Elimination of Flame- and Mechanical Straightening

CMT-Twin Welding in Automation – Application Heavy Goods Vehicle (HGV) Carriers

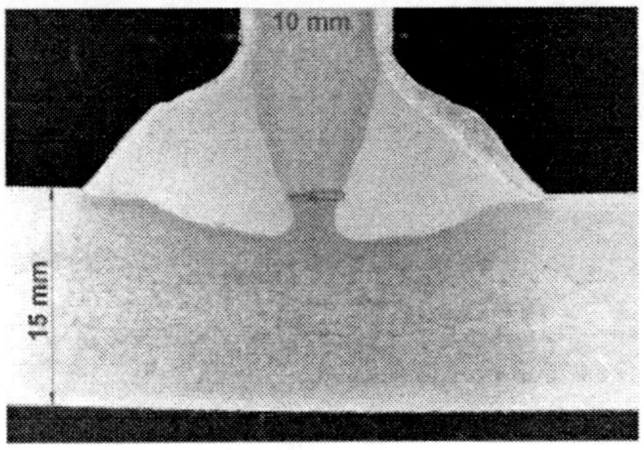

- Weld Travel Speed - t = 10 mm: 120 cm min^{-1}
- Weld Travel Speed - t = 15 mm: ≈ 250 cm min^{-1}
- Considerably reduced rework (e.g. less spatter)

CMT-Twin Welding in Automation – Application Air Pressure Vessel (Railway Industry)

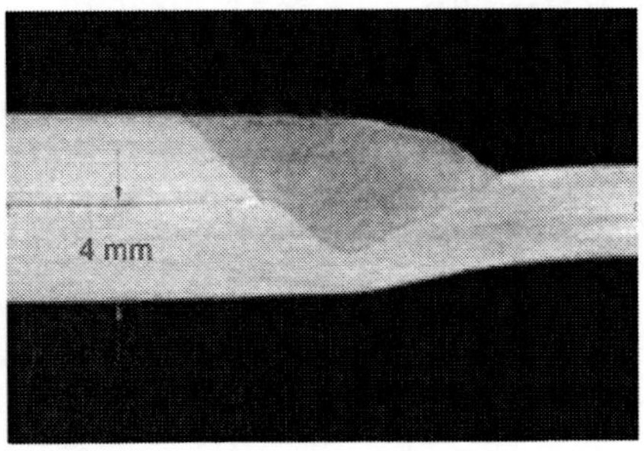

- Weld Travel Speed: 70 cm min^{-1} – 250 cm min^{-1}
- Excellent weld quality fully complying with codes and standards

CMT-Twin Welding in Automation – Air Pressure Vessel Welding: Process Validation

Conclusions

- Advanced welding automation proves one reasonable means to face the severe shortage of qualified welding personnel.
- The welding process is known playing a fundamental part in successful industrial automation systems.
- Controlled arc welding processes (CMT), by replacing e.g. GTAW, are estimated increasingly important in future welding automation.
- CMT-Twin, capable of pairing either CMT or combining CMT with distinct welding processes e.g. GMAW-P, remarkably rises process efficiency and weld quality.
- CMT-Twin proves one further quantum leap to considerably improve welding automation.

Automation of welding in fabrication sectors – An Indian experience

A. Raja, R. Easwaran and N. Rajasekaran, Tiruchirappalli/IN

The Indian welding industries have been dominantly using manual welding equipment and technologies. However, with globalisation,this situation is fast changing in every sector.Adoption of semi-automatic and automatic welding systems has been rising in recent times. The increased government spending on infrastructure like construction, energy, ports and shipbuilding, setting up of plants in India by foreign automobile companies and also increased FDI in oil & gas sector has given a boost to the technology absorption. The Indian welding/fabrication industry today is well matured to adopt any state of art welding equipment/technologies with little service support from the OEM or their Indian counterparts. The Indian welding industry is very enthusiastic to offer world class products and services. In this report, the technology scenarios of the Indian welding industries pertaining to different sectors viz. Power, Automobile, Shipbuilding, Railways and Construction have been discussed in some detail and future technology needs and potential that exist for growth are highlighted.

1. Introduction

The factorsthat drive welding automation in Indian industries may not be the same as that of the developed countries. However, the demand for improved quality at an overall lower cost, shortage of skilled manpower in certain areas such as thick wall pipe welding, sometimes even mere size and shape of the components themselves (dictating automation) could be seen as the contributing factors. By and large, quality and productivity are the watch-words of the Indian fabrication industries. Particularly, productivity is under continued pressure from intense global competition. Automation of welding has become a natural choice to meet the above challenges of quality and productivity. As a result, India's welding related market has been growing rapidly in recent years as is evident from the growing consumption of steel [Fig. 1] – India being the third biggest consumer of steel after China and the USA. The Indian market is expected to grow further with the expansion programme of energy, shipbuilding, civil construction and transportation viz. automobiles and railways sectors.

The Indian welding market has been dominated by manual welding processes and technologies for quite a long time. However, this situation is fast changing as several end-user industries have started switching over to semi-automatic and automatic welding equipments/technologies to meet the demands of improved quality and productivity. The level of welding automation in terms of volume and complexity of the technology involved may vary from sector to sector.Welding automation, at least in volume, may be the highest in Automobile followed by Power, Oil & Gas, Ship building, Construction and other sectors.

At the recently concluded Automotive Engineering Show at Chennai Trade Centre,Mr.Vijay Paranjape, President of Automation Industry Association and Director and Member Managing Board, Siemens Limited, had said: "Industrial Automation in India is in its nascent stage. Since the real drive for automation started only in early 1990s, we are still trying to play 'catch-up' with the absorption of product technologies and manufacturing processes, with geographies that have had over 100 years to learn and deploy automation. China has rapidly closed the gap in skills development and industrial deployment of highly automated processes".Perhaps, this nicely sums up the scenario!

It is attempted to give an overview of the present scenario ofthe level of welding automation and future potential in different manufacturing sectors of India.

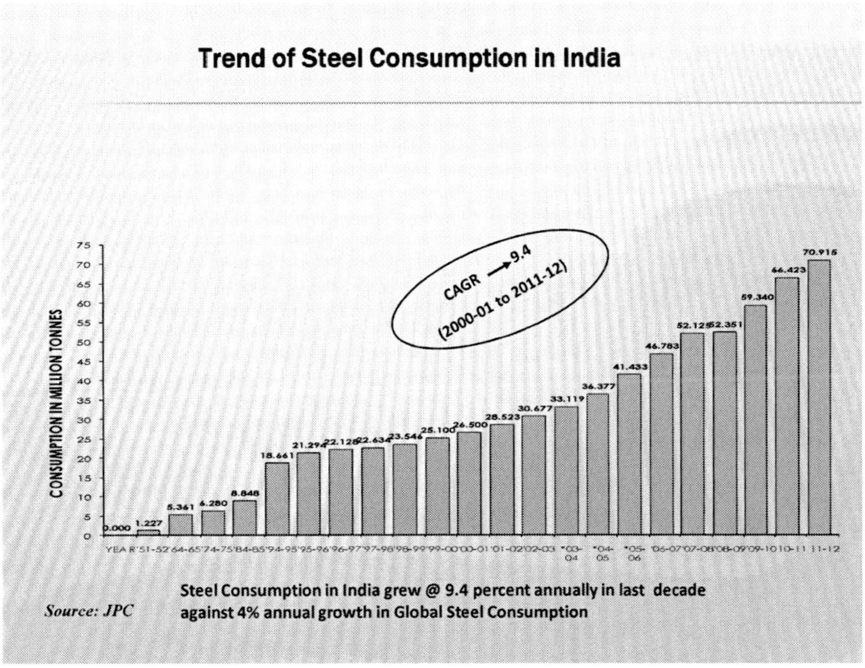

Fig1. Trend of steel consumption in India

2. Power Sector

The total installed capacity at the end of the 11th plan [31st March 2012] stands at 199877 MW with a break-up of thermal 65.84%, hydro 19.51%, nuclear 2.39% and renewable sources 12.26%. It is planned to add another 88537 MW [thermal 72340 MW, hydro 10897 MW and nuclear 5300 MW] during the 12th plan. The technologies targeted for the newer thermal plants planned in the 12th plan would be of large capacity super critical boilers and turbines of 800 MW and more [1]. Similarly ultra high voltage transmission equipments and systems are also targeted for development. This would require advanced welding technologies, equipments and systems to accomplish the task.

Manufacturing a complete Boiler for thermal power plant involves manufacture of Boiler drum,separator, header and piping, water wall panels, super heater coils and so on. The job configuration, welding position, wall thickness, type of weld and material used dictates the welding process selection.Appreciable level of automation has been achieved in the welding of Boiler drum, water wall panels and super heater coils. For example, mechanized submerged arc weldingis well established [Fig. 2] for welding of long and cir-seam joints of the boiler drum.Fixed Automation systems are often called as Special Purpose Machines [SPM].SPMs integrated with narrow gap SAW are successfully used in few industries.In order to have longer length tubes which are required to form the water wall panel, the standard tubes coming from the tube plants are joined by welding. These butt joints are carried out in SPM's which are equipped with tube feeding conveyors and self-aligning chucks and mechanized welding units. Pulsed Gas Metal Arc Welding with solid wire and Argon+CO_2 mixed gas for shielding are employed for this application. The first pass is welded without arc oscillation, the subsequent passes, which may be two or three depending on the thickness, are welded with arc oscillation. The complete arc action on the groove, i.e., the parameters for each pass, the chuck speed and crater fill parameters etc., are allpre-programmed in the system. The manufacturing practice of membrane type water wall panels has undergone sea change over the years. Traditionally, SPMs with multi-torch submerged arc welding process have been used for these joints. Recently,

FCAW process with cored wire and CO_2 gas shielding has replaced SAW in this application.SPMs integrated with Pulsed GMAW process,with solid wire and Argon+CO_2 shielding gas mixture, is also used for welding of the panels. The advantage with the pulsed GMAW is that it enables simultaneouswelding from both top and bottom of the water wall panel and this helps in avoiding flipping of the panel for backside welding [2].

Fig 2. SPM for welding Boiler drum long seam joint

There are still some areas where automation has not been successful so far. For instance, the drum nozzle joints are still welded manually using SMAW process thoughit is believed that Japan, Germany and other developed countries have adopted mechanized/robot welding for the drum nozzle joints. Attempts made to establish the drum nozzle welding with the imported nozzle welding equipments were not successful. As setting up the equipment and edge preparation was taking considerably longer time, it did not gain acceptance of the technicians/users. Similarly the headers, stubs to header and other pipe joints are still welded using manual TIG for root pass and SMAW for filler passes. Indian boiler industries are looking for an automated robust solution to weld these joints which are very critical.The added challenge is that most of these headers are with long bent stubs. Steam separator is another thick walled Grade 91 steel component in super critical boilers with multiple stubs, some of them aligned at an angle to the vertical. This is another area certainly looking for automation solutions.

Amongst the equipments that are used for thermal power generation, Steam Turbine is one of the most critical components that operate under high temperatures and pressures in extremely aggressive environments. Steam turbine components that work at high temperatures and pressures like turbine casing and steam chest that are statically loaded and other critical components like rotor, disc and blades that are subjected to creep and fatigue loading conditions. Depending on the appropriate operating temperatures, the rotor shafts can be fabricated out of low / high alloy heat resistance material or of low alloyed, highly toughened material for low pressure application. Traditionally, for the casings mono-block forging or castings made of creep resisting low alloy steels of Cr-Mo type are used while creep resistant 12%Cr type high alloy steel forging is used for blades. These forgings are finish machined as per drawings and assembled after dynamic balancing and other tests. However, the development of Super Critical and AUSC boiler technology puts a demand for development/manufacture of welded rotor shaft. Hot Wire Narrow Gap TIG Welding process has been identified for this application. Though rotor welding technology is well established in Japan, Germany and other developed countries, it is yet to be developed and applied in India [2].

Most of the automated welding systems that are used in power sector are SPM based.Robot based automated welding systems are very few. Robot welding technology is yet to be established for thick walled components involving multi-pass welding.

High pressure valves are very critical to the functioning of the thermal power plant. Manufacturing valves involve intricate welding due to the configuration of the job. All butt joints are welded with root TIG and filler passes by SMAW. Semi-automatic GMAW/FCAW is also being attempted.Seat ring to valve body welding generally called as seal welding is highly inaccessible for manual welding as the joint is located well inside the narrow bore. Besides, the hot condition of the job due to preheat make it more difficult for the welder. Some valves manufacturers have automated the seat ring to valve body welding using GMAW integrated SPMs. Gates and seat rings require stellite cladding / hardfacing to confer abrasion/corrosion resistance to these parts in service. Plasma stelliting systems/SPMs integrated with plasma torch, power source, powder feeding and AVC control is well established in valves manufacturing industries. Robotic welding automation of body yoke assembly is being attempted. There is great scope for further welding automation in valves manufacture.

Sites where the boiler and other sub-systems are erected involve welding of large number of tube and pipe butt joints in various spatial positions. The welding of these tube and pipe butt joints is mostly done by manual TIG and SMAW processes. It is attempted to introduce orbital TIG for tube butt joints and orbital FCAW for high wall thickness pipe butt joints- though one or two have already implemented orbital welding in a small way. Lack of confidence on the orbital welding process for high wall thickness large diameter pipes, apprehension about the suitability of the process in the site condition, high initial investment of the equipment, lack of adequate service support for imported equipments and non-availability of indigenous orbital pipe welding equipment are the causes for the delay in introducing this highly productive technology in the power plant erection sites, although the volume of business potential is enormous.Besides the tubes and pipes, large tonnage of structures is also welded at the erection sites. Though SMAW is used widely, semiautomatic GMAW and mechanized SAW/ GMAW also find a place in this area. There is great scope for increasing the level of automation in welding of structures also.

3. Automotive Sector

In India, automotive sectoris one where many industries have shown impressive growth over the years and has been significantly making increasing contribution to overall industrial development in the country. Presently, India is the world's second largest manufacturer of two wheelers, fifth largest manufacturer of commercial vehicles as well as largest manufacturer of tractors. It is the fourth largest passenger car market in Asia as well as a home to the largest motor cycle manufacturer.

The Indian Automobile Industry is manufacturing over 20 million vehicles [Table 1] and exporting about 3.0 million every year [Table 2]. The dominant products of the industry are two wheelers with a market share of over 75% and passenger cars with a market share of about 16%in the domestic market. Commercial vehicles and three wheelers share about 9% of the market between them. About 91% of the vehicles sold are used by households and only about 9% for commercial purposes. The industry has attained a turnover of more than USD 46 billion [Table 3] and provides direct and indirect employment to over 13 million people [3].

Table 1. Automobile Production in India

Automobile Production Trends Number of vehicles							
Category	2005-06	2006-07	2007-08	2008-09	2009-10	2010-11	2011-12
Passenger Vehicles	1,309,300	1,545,223	1,777,583	1,838,593	2,357,411	2,982,772	3,123,528
Commercial Vehicles	391,083	519,982	549,006	416,870	567,556	760,735	911,574
Three Wheelers	434,423	556,126	500,660	497,020	619,194	799,553	877,711
Two Wheelers	7,608,697	8,466,666	8,026,681	8,419,792	10,512,903	13,349,349	15,453,619
Grand Total	**9,743,503**	**11,087,997**	**10,853,930**	**11,172,275**	**14,057,064**	**17,892,409**	**20,366,432**

Source: Society of Indian Automotive Manufacturing (SIAM)

Table 2. Automobile Exports from India

Automobile Exports Trends Number of Vehicles							
Category	2005-06	2006-07	2007-08	2008-09	2009-10	2010-11	2011-12
Passenger Vehicles	175,572	198,452	218,401	335,729	446,145	444,326	507,318
Commercial Vehicles	40,600	49,537	58,994	42,625	45,009	74,043	92,663
Three Wheelers	76,881	143,896	141,225	148,066	173,214	269,968	362,876
Two Wheelers	513,169	619,644	819,713	1,004,174	1,140,058	1,531,619	1,947,198
Grand Total	**806,222**	**1,011,529**	**1,238,333**	**1,530,594**	**1,804,426**	**2,319,956**	**2,910,055**

Source: Society of Indian Automotive Manufacturing (SIAM)

Table 3. GROSS TURNOVER OF THE AUTOMOBILE MANUFACTURERS IN INDIA (IN USD MILLION)

	2006-07	2007-08	2008-09	2009-10	2010-11
	30,476	36,612	33,250	43,296	58,583
(USD Conversion Rate)	45	40	46	47	46

GROSS TURNOVER OF THE AUTOMOBILE INDUSTRY IN INDIA
2006-07 TO 2010-11

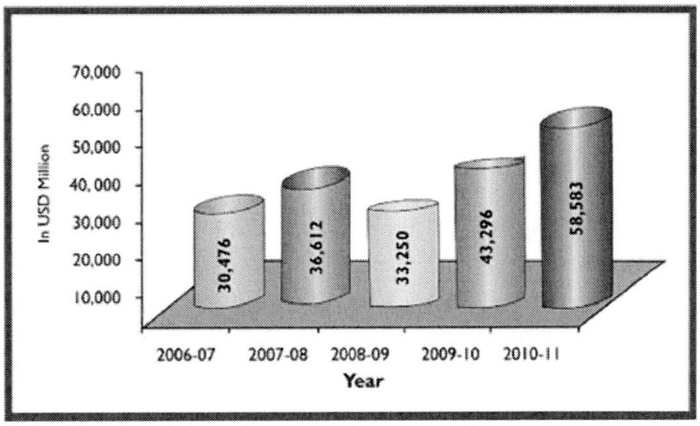

Source: Society of Indian Automotive Manufacturing (SIAM)

The Indian Automotive Industry, after de-licensing in July 1991, has grown at a spectacular rate on an average of 17% for the last few years. Automobile industry is currently contributing about 5% of the total GDP of India. India's current GDP is about USD 650 billion and is expected to grow to USD 1,390 billion by 2016. The projected size in 2016 of the Indian automotive industry varies between USD 122 billion and USD 159 billion including USD 35 billion in exports. This translates into a contribution of 10% to 11% towards India's GDP by 2016, which is more than double the current contribution [4].

The installed capacity of the automobile sector has been 15.32 million vehicles, comprising 3.17 million four wheelers (including passenger cars) and 12.15 million two and three wheelers. The sector has shown great advances in terms of development, spread, absorption of newer technologies and flexibility in the wake of changing business scenario.

Automotive sector needs speed, consistency and volume of production to be in the forefront. Therefore, automotive sector has always been the leader in adopting mechanized/automated welding technologies for fabricating the various components. Fixed automation or SPM based equipments have been and still continue to be in use in many automotive industries. After the introduction of robot welding systems, there has been a significant shift from SPM to robot welding, as it facilitates quicker change over from one model to another, besides providing higher productivity and quality. Today, the largest application of robots is in the automotive industries in India. The welding processes employed in robot welding are Gas Metal Arc Welding [GMAW] and Resistance Spot/ projection welding.High performance variants of Gas Metal Arc welding such as wave form controlled short circuiting transfer process [Lincoln STT, Miller RMD etc.], Forced short circuiting mode of welding [Rapid Arc mode] and pulse on pulse mode of welding are used by leading automobile manufacturers abroad to improve quality, productivity and to reduce spatter. Energy efficient inverter-based synergic pulsed GMA welding equipments are now widely used by automobile manufacturers in India. The usage of the high performance variants of GMAW mentioned earlier is picking up.

4. Shipbuilding Sector

Shipbuilding industry is a technology, skill and material intensive assembly operation. In the post World War II era, for nearly three decades the shipbuilding industry was dominated by Europe and the US. Shipbuilding being a labour intensive industry, the cost of labour plays an important determinant in a country's competitive position vis-à-vis others. With rising labour cost, shipbuilding activities have slowly moved away from 'high wage' Europe and US to low-wage Asia. Over the years, the production base has shifted from Europe to Japan and then to Republic of Korea and has now taken firm roots in People's Republic of China. The factors governing this shift have been relatively high wage in Europe coupled with strong competitive strength of manufacturing and steel making sectors and active state support in the leading shipbuilding nations of Asia. This decisive shift in shipbuilding activities from Europe to Asia has opened up window of o, for Indian ship yards in both public and private sectors.

Indian peninsula has a coastline of about 7,517 Kms, with 13 Major ports and 176 notified non-major Ports facilitating sea-borne trade. Coastal and overseas cargo movementis effected through ocean-going vessels. In addition, small ships/crafts also ply on inlandwaterways and canals. It is reported that approximately 95 per cent of the India's trade by volume, and 70 per cent by value, is moved through maritime transport. India is among the top 20 leading countries having large number of merchant fleets in the world. Indian owned ships/vessels carried 7.95% of India's overseas tradeduring 2010-11 and the share of Indian owned vessels in India's overseas trade during 2011-12is 10.4%. As on 31st December 2012, Indian shipping tonnage was 10.4 million GrossRegistered Tonnage (GRT), with 1154 ships. With the opening up of Indian economy toglobalization, there has been a steady increase in handling of cargo traffic at Indian Ports. Therobust growth in India's export over the last few years has improved India's share in globalexports from 0.5% in 1990 to 1.5% in 2010. To sustain the momentum of foreign trade andimprove competitiveness, the country would need adequate and efficient infrastructure in termsof ports, ships and maritime services. Further, it is targeted to achieve 5% share of the global shipbuilding market and a share of 10% in global ship repair industry by 2020 [5].

The Indian shipbuilding is mainly centered around 21 shipyards comprising 8 public sector and 13 private sector shipyards [Table 3].The shipyards have 20 dry-docks and 40 slipways between them with an estimated total capacity of 1080951 DWT.Cochin Shipyard Limited (CSL) has ashipbuilding capacity of 110,000 DWT, Hindustan Shipyard Limited (HSL), Visakhapatnam has a shipbuilding capacity of 70,000 DWT, Pipavav has a shipbuilding capacity of 400000 DWT and L&T has a shipbuilding capacity of 300000 DWT. These major public and private sector shipyards have the capacity and capability to manufacture large conventional vessels such as tankers and bulk carriers. Other domestic yards are engaged in construction of smaller vessels. Indian order book as on March 2012 was 3653010 DWT [6].

Welding is of utmost concern in shipbuilding, since several kilometres of welding is done to build a ship. SMA welding process is widely used and the usage semi-automatic processes such as solid wire GMA welding and flux cored arc welding [FCAW] isgradually increasing in the shipyards. The high productive electro-gas welding process, which facilitates vertical welding of assembly in-situ, is used by few. Development of simplified self-guided, electro-gas welding systemwill enhance the application of this process in Indianshipbuilding industries. Mechanised welding using linear tracks/carriages integrated with GMAW and FCAW are widely used across the shipbuilding industries to improve the quality and productivity. World class CNC controlled underwater plasma cutting systems are employed for material preparation.

5. Construction Sector

The Indian construction sector is witnessing a rapid growth and development in the last few years. Urban infrastructures such as metros, airports, railways, roads and ports are getting modernized. The construction component of work involved in such projectsis comparatively large: 75% in Railways, 65% in water supply, 60% in roads and 40-50% for metros, airports and ports. In industrial projects, including power, the construction component comprises only 15-20% [7, 8].

In India, the use of structural steel has been growing and it has become an important inputmaterial for construction. The higher strength, ductility, shear strength of structural steels conferring all desirable properties to the steel structure fabricated and the reduced overall cost through long service life are the reasons for preferring structural steels. High strength low alloy steels and grain refined micro alloyed steels are increasingly used to reduce weights of structures and also to achieve economy. Weather resistant steels are preferred as construction material in coastal areas. Fire resistant steels are used in important structural members (column and Beams) of buildings and this can decrease or eliminate the need for fire insulation coating. Galvanized steels and pre-painted steels are also used widely for aesthetic look.

Welding is very essential in fabrication and erection of steel structures, particularly in case of heavily loaded site connection, rigid connection and even inrefurbishment jobs. The overall economy may be achieved through the selection of right type of weldjoint, edge preparation, selection of welding process and of filler materials. Traditionally,SMAW process is employed for welding of steel structures in construction. The semi-automatic Flux Cored Arc Welding [FCAW] and Gas Metal Arc Welding [GMAW] are gradually replacing SMAW process, particularly for welding of high strength structural steels. There is a good possibility to introduce welding automation in the pre-fabrication of structural modules.

6. Railways

Indian Railways play a vital role in the economic development of the country. The Indian Railways has grown in size and strength over the years. The number of passenger trains has increased from 8897 in 2001-02 to 12335 in 2011-12.The Indian Railways have 9 major units located at various parts of the country to manufacture rail coaches, wagons and also to do repair. A new coach manufacturing unit at Haryana and two repair & rehabilitation workshops - one in Andhra and one in Odisha- are included in the 12thplan and accordingly,in the 12th plan, around Rs. 5.19 trillionis earmarked for the development of railways [9]. The nine major manufacturing units under the Ministry of Railways are well equipped with state of art equipments and facilities and they have capability to manufacture world class coaches, wagons and other items required by the railways. With regard to welding, the high productive semi-automatic GMAW/FCAW is widely practiced. GMAW integrated SPMs and gantry robot automated welding systems are widely employed for the fabrication of the sub-assemblies of the under-frame. Fully automated large capacity Flash Butt welding systems are used for joining the rails. However, Railways is looking for innovative welding processes for field welding of rails and frog joints. Good reclamation technology for rebuildingthe worn-out rails in-situ is also sought after.

Metro rail is another segment of importance in India. In order to decongest the traffic in the cities, major Indian cities like Delhi, Calcutta, Bangalore and Chennai are opting for metro trains as viable alternative.The Delhi Metro currently operates across 156 km,making it one of the largest networks in the world. Both public sector and private majors are involved in the manufacture of these stainless steel coaches. Robotic resistance spot welding technology is well established and used in these manufacturing units for the fabrication of the stainless steel coaches for the metros.The lap joints in the roof are welded by manual TIG welding process. In the robotic resistance spot welding, the manufacturers face occasional lack of bonding problem due to variation in material condition and other process settings.Robust process controller is being considered for online monitoring of the process, ensuring corrective action during robotic resistance spot welding to avoid the above issue.

7. Review of welding automation technology

SMAW has been the prime welding process used in all the fabrication industries for many years. But there has been a shift from SMAW to high productive semi-automatic processes such GMAW, FCAW & SAW in the recent past. The introduction of energy efficient and dynamically superior inverter power sources also helped in improving the field performance of these welding processes. The continuous wire processes also favoured welding automation by their very design. Programmable variable speed carriages integrated with GMAW/SAW head and equipped with or without tactile sensors can be seen in many fabrication industries for fabrication of large columns, sub-assemblies of higher capacity transformers etc. In the automotive sector, involving single pass fillet, lap joints in thin walled tubes and sheets, robotic welding employing GMAW process is successfully established. However, in heavy engineering applications involving thick sections requiring multi-pass welding, mostly SPMs are used and robotic welding automation is yet to be established this field. The SPMs are viewed as less complicated and easy to handle by the technician. Nevertheless, the success of robotic welding in such applications in the developed countries provides a kind of encouragement to the Indian industries to move in that direction.

Laser and hybrid-laser GMA welding are state of art technologies. International automotive majors like Volkswagen who have set up their manufacturing plant in India, has adopted automated high productive processes like robotic laser joining for fabrication of automobile components. Taking cue from such developments, various Indian automotive industries are also attempting laser welding/joining for their components especially for Tailor Welded Blanks (TWBs).

8. Conclusions

The various sectors of the Indian economy have progressed well thus far and ambitious growth plan of each sector will steer the future growth. The fabrication industry, particularly the welding industry, catering to these different sectors has reached a stage of maturity to absorb/adopt any latest state of art technology. The success with the deployment of state of art automated welding technologies will largely depend on the continued service support from the OEM or their Indian counterpart. In any case, welding automation will becomeinevitable in the given circumstances and be a natural choice for Indian industries to remain competitive and deliver world class product and services. The potential for welding automation either robotic or otherwise remain enormous in India.

References

1. Power scenario at a glance; CEA Report, November 2012.
2. Dr. A. Raja: Automation in welding of power plant components. IWS 2k10 International Welding Symposium on "Innovative processes for Material joining & Surfacing", 10th - 12th February 2010, Mumbai.
3. Automobile Industry India - Executive summary. www.imaginmor.com
4. Automotive Mission Plan 2006–2016: Ministry of Heavy Industries & Public Enterprises – Report.
5. The manufacturing plan: Strategies for accelerating growth of manufacturing in India in the 12th Five Year Plan and Beyond, Report of Planning Commission, Govt. of India.
6. Statistics of India's shipbuilding and ship repair industry 2011-12.Ministry of Road Transport and Highways, Transport Research Wing, New Delhi – Report.
7. Anandrathi: India construction sector - breaking the gridlock. Construction Sector Report – 9th March 2012.
8. Dr. Jayanta K. Saha: Welding options in steel construction. Institute for Steel Development & Growth, Kolkata,India – Report.
9. Highlight of the Railway Budget 2013-14. Ministry of Railways – Report.
10. Made in India Metro Rolls out – Times of India, 6th June 2009.

Research and development of arc welding processes for automation in welding in Japan

Y. Hirata, Osaka/JP

Global and domestic competition related with welding and metal construction field is getting severe in Asia region. Then, low cost production with high weld quality is strongly demanded. These years, various fundamental researches have been conducted in not only universities/institutes but industries in order to develop high productivity welding and/or fully mechanized/robotic welding in Japan.

GMAW process is widely used in various production fields because it provides high deposition rate and/or it is suitable for automatic welding and robotic welding. The series of wire melting and subsequent molten drop transfer process makes deposition metal of butt weld and/or fillet weld. So, the stability of gas metal arc with metal transfer dominates the weld qualities and productivities. These technical issues have been promoting welding machine makers, consumable makers, gas-suppliers to improve and develop high performance welding processes. In this paper, current technologies related mainly with GMAW processes oriented to automatic welding in Japan are described.

AUTHOR INDEX

International Institute of Welding
BP 51362 – VILLEPINTE
95 942 ROISSY CH DE
GAULLE Cedex
France

ISBN 978-1-63439-610-3
9 781634 396103

CURRAN ASSOCIATES INC.
proceedings.com

Annual Conference and Expo of the Association of Avian Veterinarians (AAV 2025)

Phoenix, Arizona, USA
3-6 October 2025

ASSOCIATION OF AVIAN VETERINARIANS
AAV